CAMBRIDGE COUNTY GEOGRAPHIES

General Editor: F. H. H. GUILLEMARD, M.A., M.D.

HEREFORDSHIRE

CAMBRIDGE UNIVERSITY PRESS
Cambridge, New York, Melbourne, Madrid, Cape Town,
Singapore, São Paulo, Delhi, Mexico City

Cambridge University Press
The Edinburgh Building, Cambridge CB2 8RU, UK

Published in the United States of America by Cambridge University Press, New York

www.cambridge.org
Information on this title: www.cambridge.org/9781107678866

First published 1913
First paperback edition 2013

A catalogue record for this publication is available from the British Library

ISBN 978-1-107-67886-6 Paperback

Cambridge County Geographies

HEREFORDSHIRE

by

A. G. BRADLEY

With Maps, Diagrams and Illustrations

Cambridge:
at the University Press
1913

PREFACE

I SHOULD like to take this opportunity of thanking the many residents in the county of Hereford who have kindly answered queries or verified statements of detail contained in this book. Special thanks are due to the Rev. Hubert Mallinson who contributed several photographs especially taken for it, to the Rev. Canon Bannister for looking over portions of the proofsheets, to the late Rev. Augustin Ley in the matter of the botanical notes, and to Mr T. Hutchinson in the paragraphs dealing with ornithology. Lastly I wish to acknowledge the kindly aid of Mr S. C. Roberts of the University Press in the general revision of the proofs.

A. G. B.

January, 1913.

CONTENTS

		PAGE
1.	County and Shire. The word *Hereford*. Its Origin and Meaning	1
2.	General Characteristics. Position and Natural Conditions	4
3.	Size. Shape. Boundaries	10
4.	Surface and General Features	13
5.	Watersheds and Rivers	19
6.	Geology	24
7.	Natural History	31
8.	Climate and Rainfall	38
9.	People—Race, Dialect, and Population	44
10.	Agriculture	54
11.	Industries and Manufactures	61
12.	History of the County	65
13.	Antiquities—Prehistoric, Roman, Saxon	77
14.	Architecture—(*a*) Ecclesiastical	83

CONTENTS

PAGE

15. Architecture—(*b*) Military 98
16. Architecture—(*c*) Domestic 103
17. Communications—Past and Present. Roads and Railways 114
18. Administration and Divisions—Ancient and Modern . 119
19. Roll of Honour 123
20. The Chief Towns and Villages of Herefordshire . 131

ILLUSTRATIONS

	PAGE
Hereford Cathedral	3
View of the Wye from Sugwas Cliff	7
A Herefordshire Hop Garden	8
The Wye at Hay	12
Grwyny Fawr	13
The Malvern Hills	14
Symond's Yat	16
Bromyard Downs	17
The Wye at Hereford	19
The Wye at Kerne Bridge	20
Horseshoe bend of the Wye at Ross	21
The Junction of the Monnow and Honddu	22
Coracles on the Wye	35
Seven Sisters Rocks, on the Wye	37
Portion of Domesday Book relating to Herefordshire	51
A Cider Orchard	57
Hereford Bull	58
Cattle Market, Ledbury	60
Cider Works, Hereford	62
Site of St Ethelbert's Well, Castle Hill, Hereford	68
Castle Tump	70
Memorial Stone on the battle-field of Mortimer's Cross	74
The Herefordshire Beacon	78
Arthur's Stone	80
Norman Nave, Leominster Church	86
Kilpeck Church: Norman Doorway	88
Ducking Stool, Leominster Church	89

PAGE
Abbey Dore Church 90
Bromyard Church 91
Pembridge Church 92
Ledbury Church 93
Colwall Church 95
Ross Church 97
Goodrich Castle 100
Wilton Castle 101
Bollitree Castle 102
Old house, Hereford 104
Bargates, Leominster 105
Blackfriars Monastery, Hereford 109
Market House, Ledbury 112
Market House, Ross 113
Eardisland Dovecote 114
Wilton Bridge, near Ross 118
Hereford Cathedral School 122
Nell Gwynne 125
David Garrick 127
David Cox 129
The Choir: Hereford Cathedral 135
The Hereford Mappa Mundi 136
Church House, Ledbury 139
Leominster Church 141
Grange Court House, Leominster 142
Pembridge 144
Diagrams 147

MAPS

England and Wales, showing Annual Rainfall . . 41
Chief Castles of Wales and the Border Counties . *facing* 98

The illustrations on pp. 22, 70, 74, 80 and 92 are from photographs by the Rev. Hubert Mallinson; those on pp. 3, 7, 16, 17, 19, 20, 21, 35, 37, 68, 88, 89, 90, 91, 93, 97, 100, 101, 102, 104, 105, 109, 112, 113, 118, 135, 139, 141 and 142 by Messrs F. Frith and Co.; those on pp. 8, 14, 57, 62, 78 and 122 by Mr L. J. Starkey, Hereford; those on pp. 13, 86, 114 and 144 by Mr W. M. Dodson, Bettws-y-Coed; those on pp. 125, 127 and 129 are from photographs by Mr Emery Walker of portraits in the National Gallery; those on pp. 58 and 136 are from the *Encyclopaedia Britannica* (11th Edition); that on p. 12 is from a photograph by Mr P. B. Abery, Builth Wells; that on p. 51 by Mr R. B. Fleming; those on pp. 60 and 95 by Messrs J. Valentine and Sons. The map facing p. 98 is from a drawing by Mr C. J. Evans.

1. County and Shire. The word *Hereford*. Its Origin and Meaning.

In considering the names of the counties of England, the first point that strikes us is that while most of them have the ending "shire," others, like Kent and Norfolk, are without it. Of the latter it may be generally assumed that they are the survivals of old English kingdoms or of independent communities; thus Sussex and Essex represent the kingdoms of the South and East Saxons respectively, while Northumberland recalls the ancient realm of Northumbria. The suffix *shire*, on the other hand, indicates one of the "shares" or parts "shorn off" —for the words are of similar meaning—into which the more developed Saxon kingdoms were afterwards divided.

Herefordshire was the south-westerly *shire* of Mercia, the kingdom of the Marches. The name was taken from the town of Hereford, on the Wye, which had already attained some importance. As to the derivation of the word, various conjectures have been made: the most common is that which derives it from the Welsh Hênffordd—"the old road." Leland, the famous antiquary of the sixteenth century, says of the town that it "standeth somewhat lowe on every syde. The name of it by some in Welsh is called *Heneford*, of an old ford by the castle, or ever the great bridge on the Wye was made." Others, again, derive it from Saxon terms meaning the *Ford of the Army*. But all these are merely conjectures and the most that is definitely known is that the name of Hereford, originally that of a town on the Wye, was transferred to the whole shire when the Mercian kingdom was split up. In any case, no such importance belongs to the name as in the case of those derived from the names of communities, Norfolk, for instance (the North folk of the kingdom of East Anglia), or Wiltshire (the clan country of the Saxon Wilsoetas). On the whole, a Welsh derivation of *Hereford* is more probable than a Saxon one; first, because the district west of the Severn was not won from the Welsh by Offa, king of Mercia, until the latter half of the eighth century, and secondly, because, as we shall see, people of Welsh blood and speech remained in permanent possession of a large portion of the county as contented and incorporated members of the Saxon community.

The actual county was formed at the end of the Saxon period, and Hereford became the centre of a diocese which roughly included the whole of Mercia west of the Severn. The Domesday Survey of 1086—which contains an interesting description of how "in the city of Hereford there were 103 men dwelling together within and without the wall," who had certain prescribed customs—included Caerleon and Monmouth, which had been recently acquired from the Welsh, within the boundaries of *Herefordscire*. On the other hand, the county was smaller on the north-western border than

Hereford Cathedral

it is now. In later times the chief readjustment of boundary was due to the abolition of the Marches of Wales and their conversion into counties during the time of Henry VIII.

2. General Characteristics. Position and Natural Conditions.

It is not always easy for us, to whom Great Britain has for centuries been an accomplished fact, to realise the importance which belonged, in the past, to the "Border counties" of England. Of these there were six, Northumberland and Cumberland on the Scottish border, and Cheshire, Shropshire, Herefordshire, and Monmouthshire (created in 1536) on the frontier of Wales. From the nature of their past history, and from the fact of their having been for many centuries on constant guard against a formidable and warlike neighbour, these counties have a peculiar character. Herefordshire, it is true, has not been the scene of so many great historical events and pageants of a purely political and peaceful order as many of the larger, more populous, and wealthier counties of England; but in the matter of battles and sieges, and all the incidents associated with constant invasion by hostile races, it is much richer than counties that have never been on guard against nimble enemies almost sleepless in their hostility, like the Welsh and Scots.

Herefordshire therefore, though smaller, poorer, and less populous than many other counties, will always possess to a high degree that peculiar interest which belongs to a border county. And of what eventful and stirring a nature this interest is will be shown on a later page. In another sense too, and a physical one, Herefordshire gains greatly from being a border county, as it abuts on the boundaries of South Wales, and not merely possesses the lofty and shapely range of the Black Mountains as an immediate background to much of its landscape, but actually includes their outer barriers within its own limits. Fortunate in this respect upon the west, it is only less so on the east, where the long lofty wall of the Malvern Hills occupies very much the same protecting and dominant attitude, and also stands partly within the county boundary. Though the Black Mountains are real mountains, as such are reckoned in Great Britain, here and there almost reaching the altitude of such well-known peaks as Helvellyn and Cader Idris, and surpassing that of Plinlimmon, the Malverns on the other hand are only the highest English hills between the Channel and the Peak of Derbyshire.

But this description does the Malverns less than justice, for though the highest peak is scarcely 1400 feet above the sea, the long array of sharp summits, when seen at a distance, has a distinction quite unapproached by any range of so modest an altitude in Great Britain. They are far more like real mountains, when snow-clad, or seen against a sunset sky, or on a dark day from a little distance off, than are, for instance, the higher and wilder heights of Dartmoor or Exmoor in Devonshire. Between these high ranges Herefordshire lies as an undulating plain, throwing up everywhere isolated hills or long ridges of from six hundred to a thousand feet. These hills enjoy the rather rare and picturesque quality of being draped in rich woods to their summit, a peculiar and attractive characteristic of Herefordshire scenery. Our shire is beyond all question one of the most beautiful of English counties. For the far north of England, like most of Wales, does not admit of comparison, but belongs to another order of landscape, on a wilder, loftier, and more imposing plane.

But, for the rest, Herefordshire, to those of open mind who know it and also know England well, will generally be thought to hold its own among other inland counties in physical beauty. Monmouth particularly (if it be reckoned as English) and Devonshire will naturally suggest themselves in friendly rivalry, as similar qualities win praise from the stranger in all three. A considerable portion too of Shropshire belongs to the same class of landscape. But if Hereford has not the great moors of the famous west of England county, it has on the other hand real mountain prospects from every slight elevation within its limits, and one qualification that is perhaps unique outside Cumberland and Westmorland. For, without exception, the other counties of England distinguished for beauty of landscape include considerable stretches that are commonplace and uninteresting. But there is practically no single stretch of Herefordshire landscape that does not rise considerably above the average English standard of physical beauty. It is moreover watered by the Wye, which has the varied, rapid current and

View of the Wye from Sugwas Cliff

the clear water of a river of mountain origin, while nearly all the streams of Herefordshire rise in the Welsh hills, and flow not merely through the more mountainous western valleys of the county but through the lower and more ornate portions, and through park and pasture lands where they are shaded by the stateliest timber. The soil

A Herefordshire Hop Garden

is nearly everywhere a rich red loam, and the pastures have the peculiarly bright green tint of the western country and climate. Plant life responds to kindly conditions, not merely when cultivated by man, and ferns and wild flowers flourish everywhere profusely. Hop gardens and orchards are, of course, a leading feature in Herefordshire scenery, as are also the meek, white-faced, red Herefordshire cattle, which have still a monopoly of its pastures and hold the field against intruders.

Herefordshire is purely an agricultural county. Together with Worcestershire it forms the second great hop district of England—Kent and East Sussex being the chief one. It shares with Devonshire, too, the honour of being the predominant apple and cider district of England. Fuller, the famous chronicler of Old England, says: "This county doth share as deep as any in the alphabet of our commodities, though exceeding in W for wood, wheat, wool, and water. Besides, this shire better answereth (as to the sound thereof) the name of Pomerania than the dukedom of Germany so called, being a continued orchard of apple-trees, whereof much cider is made."

The climate is mild and equable, though there is greater moisture than is usual in England, owing to the mountainous character of the district, and the soil is, on the whole, rich and warm. The county is by no means rich in the sense of earned or accumulated wealth. The returns of agriculture in these days are small, and the value of land, Herefordshire's almost sole source of wealth, has, here as elsewhere, declined under the stress of foreign competition. It has no manufacturing or industrial life, such as is found in portions of neighbouring counties like Cheshire and Monmouth and in a less degree in Shropshire, nor does any great neighbouring city extend its wealth and population into any part of it. But nature, on the other hand, has endowed the county of Hereford with all those advantages which make for success in the improved and more intensive system of agriculture that will some day, no doubt, be more generally prevalent in England. In the meantime it is a county both good to look upon and good to live in, though with an air perhaps more pleasant than stimulating to strangers used to the vigorous breezes of the north and east.

3. Size. Shape. Boundaries.

The boundaries of our English counties vary much in character. Some, like Norfolk, Suffolk, and Cornwall are of simple outline and largely delimited by sea and river. Others, like Hereford, are extremely complex in shape, and rivers, though numerous and well-adapted to serve as convenient boundaries, have been made but little use of for this purpose.

Herefordshire in point of size stands 29th among the 41 English counties. It is about 40 miles in length measuring from north to south, about 35 miles from east to west, and has an area of about 840 square miles. Its total area in statute acres, including inland water, is officially given as 538,924 acres.

Roughly the county is an irregular oval, or blunt-pointed diamond in form, with the long axis running nearly north and south, and the county town nearly in the centre. We may begin our peregrination of the boundaries close to the southern point, just north of Monmouth, where the Wye forms the limit for some three miles or so, rounding Doward Hill and Lord's

Wood until Symond's Yat is reached, a lovely spot much visited by tourists. Here the boundary makes a leap across to the river at the base of the opposite hill, leaving a long loop of the Wye within the county. It still follows the stream for another three or four miles and then leaves it to turn eastward over the high ground towards Mitcheldean and over the northern slope of May Hill. Hence it holds a mainly northern course, varied by innumerable irregularities, for many miles, following no natural boundaries with the exception of the Malvern Hills, along the summit of which it runs till near their northern end, where it drops on the western side so as to leave the Worcester Beacon in that county. At Whitbourne the river Teme becomes the boundary for two or three miles, but the line soon leaves the low ground and proceeds by a most irregular course in a general north-westerly direction towards Ludlow, striking the Teme again in the neighbourhood of Little Hereford, where for two short stretches this river again forms the boundary. Bingewood Chase is then crossed, and about six miles west-north-west of Ludlow, near Ferney Hall, the county reaches its most northern point.

The western limits of Herefordshire are perhaps not quite so irregular as the eastern, but they show the same independence of natural boundaries, and scarcely any Hereford towns or villages of any importance are situated upon the border. At Brampton Bryan the Shropshire border is exchanged for that of Radnor and the line leads south-west over ground above 1000 feet in height to Presteign, part of which lies within our county. Still trending south, Offa's Dyke and the heights of Bradnor Hill are passed, and the boundary crosses Hergest Ridge (1389 feet) and drops down to Huntington and Michaelchurch, at which point the river Arrow enters the county, after having formed the boundary for about a mile. A few miles further south the Wye is struck at Rhydspence, and followed to Hay, where the line leaves

The Wye at Hay

it to trace the Cusop Dingle to its source and climb the Black Mountain to its summit ridge, which it follows almost to its southern end, only leaving it at Hatterall Hill to drop into the valley of the Monnow.

The Monnow forms the south-western limit of our county for many miles, with the exception of a small loop at Skenfreth, until the neighbourhood of Rockfield is reached, when the boundary line leaves it to cut across by Buckholt and Ganarew to the point from which we started.

A strip of land on the east side of the Grwyny Fawr, called the Ffwddog, belonged to Hereford till 1893, and still goes with that county for parliamentary purposes.

Grwyny Fawr

4. Surface and General Features.

From what has already been said in connection with the topography and scenery of Herefordshire it will be gathered that its surface is generally hilly, though even apart from the level meadow lands along the river valleys there are a few districts that would be more properly described as gently undulating. Taking the shape of the county to be that usually described with a little freedom as diamond, its bolder and loftier regions lie along the two westerly sides, always excepting the Malvern Hills, along the summit of whose narrow and elevated range of peaks the extreme eastern part of the county border-line runs. The north-western side, from Ludlow—which is just in

The Malvern Hills

Shropshire and stands near the top point of the imaginary diamond—to where the Wye enters Herefordshire at the western angle, is a continuous network of hills. Most conspicuous is the lofty, winding, and forested ridge from Bingewood Chase to Aymestrey, of which High Vinnals (1150 feet) is the highest point. Then come the hills clustering thick along the Radnor border and of about the same height, among which Hergest (1389 feet) and Brilley are conspicuous, being outliers of those great upland moors just behind them known as Radnor Forest, which rises to 2000 feet. South of the Wye, where the county boundary trends eastward, the Black Mountain system begins. Its most southerly main rampart, Hatterall Ridge, carries the county line and has an altitude of 2000 feet. In this neighbourhood Herefordshire touches the highest altitude (2306 feet) of any English county south of Yorkshire. Thence an array of somewhat lower parallel ranges roll northward and eastward, bearing in their several troughs the waters of the Monnow and two of its tributaries. The last two ridges, which enclose between them the celebrated Golden Valley, watered by the river Dore, are not of a semi-wild character like the others, but fertile and well occupied. Immediately without the county in the curious tooth-like projection of Monmouth which here shoots up to the northward, and between the Hatterall Ridge and the next shoulder of the Black Mountains, which again is that outlying strip of Herefordshire already alluded to, is the beautiful valley of the Honddu, which the great ruins of Llanthony Priory, standing mountain-girdled in its remote recesses, have made historic.

Towards the southerly point of the county are many shapely heights, all nearly a thousand feet high, looking down upon the twisting course of the Monnow, which here, for some distance fretting in deep valleys, chafes the feet alternately of the Monmouth and Hereford hills. Of these latter Garway is the most conspicuous.

Within these bolder ramparts of the county are many shorter ridges and many isolated hills, mostly, as before mentioned, wooded to their summits. Orcop and Aconbury, south of Hereford, are both over 900 feet. Dinedor, nearer still, is 595 feet. A range of hills runs northward with scarcely a break from just east of Ross to Stoke Edith, following a zigzag course of 20 miles and

Symond's Yat

reaching a height of nearly 900 feet at Seager Hill. Dinmore Hill is a long wooded ridge, striding the Hereford and Ludlow road at 520 feet just south of Leominster. Credenhill, Brinsop Hill, Ladylift, and Wormsley Hill are all conspicuous wooded heights to the north-west of Hereford; while of the many lower hills that crop up in picturesque fashion all over the county it is impossible here

Bromyard Downs

to take note. One remarkable bit of nature's handiwork Hereford shares equally with Gloucester in the extreme south, namely, that bold and lovely gorge famous throughout England and known as Symond's Yat, where the Wye rushes with impetuous stream through a succession of precipitous hills from 500 to 600 feet high, displaying a glorious blend of grey crags and rich and varied woodland. There are no plains save the river flats in Herefordshire, nor on the other hand is there any genuine moorland, if we except the narrow strips along the edges of the Black Mountains which overstep the county line. There are many large stretches of upland common, however, clad with bracken or heather, such as those at Ewyas Harold and about Hergest and Aymestrey, and again above Bromyard. There are no "forests" in the sense of old Royal hunting grounds or Crown properties, though the Forest of Dean touches the county about Symond's Yat.

An ancient "chase," however, belonging to Kentchurch Court, climbs almost to the top of Garway Hill, and contains yew trees old enough to have been standing when it belonged to their Scudamore owner who was Glyndwr's son-in-law. The continuously wooded and lofty ranges radiating from High Vinnals form perhaps the most extensive elevated woodland in England. It was in their recesses that the incident occurred which caused Milton, then living at Ludlow just beneath, to write the Masque of *Comus*, and to-day the only fallow deer in the kingdom that breed in a wild state roam their vast woods.

5. Watersheds and Rivers.

The Wye—by far the most important in our county—is one of the greater rivers of England and is generally admitted to be the most beautiful. It reaches Herefordshire from Wales on its western border at the Breconshire town of Hay. Thence it flows eastward to the city of Hereford,

The Wye at Hereford

nearly in the centre of the county, and from Hereford pursues a southerly course by Ross to Goodrich Castle and Kerne Bridge. Here for a few miles through the famous scenery about Symond's Yat it divides the counties of Gloucester and Hereford, leaving the latter about four miles short of the town of Monmouth, where it receives the Monnow. Though so large and important, the Wye maintains more or less the character of a mountain river throughout its whole course till it actually meets the tide. Being frequently broken by shallows and rapids, it is nowhere in Herefordshire navigable for serious commercial purposes, though in many places an ideal stream for boating. It rises close to the source of the Severn in

The Wye at Kerne Bridge

the high solitudes of Plinlimmon, a fact which is the subject of several old Welsh poems, and though the wilder and more sublime beauty which characterises its passage through Wales gives way to more placid meanderings on entering Herefordshire, its pebbly bottom, the clearness and liveliness of its waters, the picturesque homesteads and villages on its banks, the pastoral and wooded scenery it steals or ripples by, make it always an unusually attractive river. Its long passage, however, through the heart of Herefordshire, 60 miles measured by its windings, is by comparison a quiet intermediary stage between the violence with which it frets over rocky channels for the first 50 miles of its boisterous

Horseshoe bend of the Wye at Ross

journey amid deep troughs of Welsh hills and mountains, and its sinuous wanderings amid the equally beautiful, but somewhat different class of scenery through which it passes from Ross to Chepstow. Hereford and Ross are the only two places of any importance upon the Wye within the county limits.

Though all far inferior in size and the length of their

course to the Wye, the other rivers of importance in the county are the Lugg, the Arrow, the Monnow, and the Teme. All these, though comparatively small, have a reputation far beyond Herefordshire for their trout and grayling. They are quite useless for navigable purposes, for like the Wye, on a smaller scale, they maintain

The Junction of the Monnow and Honddu

throughout their course the shallow and rapid character of mountain streams. The Monnow rises in the Black Mountains just within the county limits, and from near Pandy station to close upon its junction with the Wye at Monmouth forms the county boundary, except for a small loop near Skenfreth. Early in its career it receives several unimportant brooks: at Pandy the Honddu joins it from the Llanthony valley, and at Pontrilas the Dore, which waters the Golden Valley, comes in reinforced by the Worm, a sluggish brook flowing from Allensmore near Hereford.

The Lugg rises in the northern part of Radnor Forest, enters our county at the north-west near Presteign, and flowing through deep valleys to Mortimer's Cross enters a more open country and continues its course a little south of west to Leominster. Thence turning due south it pursues its zigzag way to the Wye at Mordiford, four miles below Hereford. The Arrow also rises in Radnor Forest and, entering the county near Michaelchurch, flows by Kington, Pembridge, and Eardisland to join the Lugg some two miles below Leominster after having received the waters of the Stretford Brook.

The Teme, which enters the county at Brampton Bryan west of Leintwardine, where it is joined by the Clun from Shropshire, is a rather larger river than any of these others, but of the same type. Its six or seven mile course through the extreme northern point of the county is practically all that it has to do with Herefordshire, save for a further incursion of a mile or two near Brimfield and again for as brief a space as a boundary near Ludlow and Whitbourne. But for a portion of its short way below Leintwardine the Teme changes the character in which it entered the county, and at Downton Castle plunges for a considerable distance over rocky ledges between wooded cliffs, with all the rugged beauty and the turmoil of a river in the Welsh mountains, presenting rather a curious physical phenomenon.

All these rivers and almost all their feeders rise in the Welsh hills. The watersheds that divide their short courses before they unite in the Lugg and Wye are of small importance. The Teme alone flows to the Severn, and carries with it the waters of but a few insignificant Herefordshire brooks.

The streams of east Herefordshire are unimportant and less rapid. The Frome rising on the county border some distance north of Bromyard flows south into the Lugg near its junction with the Wye. The Leadon, running through Ledbury, leaves the county just below that town to join the Severn further south.

Speaking generally, the waters of Herefordshire run from the east or west of the county towards its centre and south, and are then carried southward by the Wye into the Bristol Channel. Scarce even a nameless brook runs north or out of Herefordshire into the Teme.

There is not a natural lake in the county. But it may be mentioned that the Wye, through the medium of its tributaries in Central Wales, provides Birmingham with a pure and abundant supply of water.

6. Geology.

By Geology we mean the study of the rocks, and we must at the outset explain that the term *rock* is used by the geologist without any reference to the hardness or compactness of the material to which the name is applied; thus he speaks of loose sand as a rock equally with a hard substance like granite.

Rocks are of two kinds, (1) those laid down mostly under water, (2) those due to the action of fire.

The first kind may be compared to sheets of paper laid one over the other. These sheets are called *beds*, and such beds are usually formed of sand (often containing pebbles), mud or clay, and limestone, or mixtures of these materials. They are laid down as flat or nearly flat sheets, but may afterwards be tilted as the result of movement of the earth's crust, just as we may tilt sheets of paper, folding them into arches and troughs, by pressing them at either end. Again, we may find the tops of the folds so produced worn away as the result of the action of rivers, glaciers, and sea-waves upon them, as we might cut off the tops of the folds of the paper with a pair of shears. This has happened with the ancient beds forming parts of the earth's crust, and we therefore often find them tilted, with the upper parts removed.

The other kinds of rocks are known as igneous rocks; these have been melted under the action of heat and become solid on cooling. When in the molten state they have been poured out at the surface as the lava of volcanoes, or have been forced into other rocks and cooled in the cracks and other places of weakness. Much material is also thrown out of volcanoes as volcanic ash and dust, and is piled up on the sides of the volcano. Such ashy material may be arranged in beds, so that it partakes to some extent of the qualities of the two great rock groups.

The relations of such beds are of great importance to geologists, for by means of them we can classify the rocks according to age. If we take two sheets of paper, and lay one on the top of the other on a table, the upper one has been laid down after the other. Similarly with two beds, the upper is also the newer, and the newer will remain on the top after earth-movements, save in very exceptional cases which need not be regarded here, and for general purposes we may look upon any bed or set of beds resting on any other in our own country as being the newer bed or set.

The movements which affect beds may occur at different times. One set of beds may be laid down flat, then thrown into folds by movement, the tops of the beds worn off, and another set of beds laid down upon the worn surface of the older beds, the edges of which will abut against the oldest of the new set of flatly deposited beds, which latter may in turn undergo disturbance and renewal of their upper portions.

Again, after the formation of the beds many changes may occur in them. They may become hardened, pebble-beds being changed into conglomerates, sands into sand-stones, muds and clays into mudstones and shales, soft deposits of lime into limestone, and loose volcanic ashes into exceedingly hard rocks. They may also become cracked, and the cracks are often very regular, running in two directions at right angles one to the other. Such cracks are known as *joints*, and the joints are very important in affecting the physical geography of a district. Then, as the result of great pressure applied sideways, the rocks may be so changed that they can be split into thin slabs, which usually, though not necessarily, split along planes

	Names of Systems	Subdivisions	Characters of Rocks
TERTIARY	Recent Pleistocene	Metal Age Deposits Neolithic ,, Palaeolithic ,, Glacial ,,	Superficial Deposits
	Pliocene	Cromer Series Weybourne Crag Chillesford and Norwicn Crags Red and Walton Crags Coralline Crag	Sands chiefly
	Miocene	Absent from Britain	
	Eocene	Fluviomarine Beds of Hampshire Bagshot Beds London Clay Oldhaven Beds, Woolwich and Reading [Groups Thanet Sands	Clays and Sands chiefly
SECONDARY	Cretaceous	Chalk Upper Greensand and Gault Lower Greensand Weald Clay Hastings Sands	Chalk at top Sandstones and Clays below
	Jurassic	Purbeck Beds Portland Beds Kimmeridge Clay Corallian Beds Oxford Clay and Kellaways Rock Cornbrash Forest Marble Great Oolite with Stonesfield Slate Inferior Oolite Lias—Upper, Middle, and Lower	Shales, Sandstones and Oolitic Limestones
	Triassic	Rhaetic Keuper Marls Keuper Sandstone Upper Bunter Sandstone Bunter Pebble Beds Lower Bunter Sandstone	Red Sandstones and Marls, Gypsum and Salt
PRIMARY	Permian	Magnesian Limestone and Sandstone Marl Slate Lower Permian Sandstone	Red Sandstones and Magnesian Limestone
	Carboniferous	Coal Measures Millstone Grit Mountain Limestone Basal Carboniferous Rocks	Sandstones, Shales and Coals at top Sandstones in middle Limestone and Shales below
	Devonian	Upper, Middle, Lower — Devonian and Old Red Sandstone	Red Sandstones, Shales, Slates and Limestones
	Silurian	Ludlow Beds Wenlock Beds Llandovery Beds	Sandstones, Shales and Thin Limestones
	Ordovician	Caradoc Beds Llandeilo Beds Arenig Beds	Shales, Slates, Sandstones and Thin Limestones
	Cambrian	Tremadoc Slates Lingula Flags Menevian Beds Harlech Grits and Llanberis Slates	Slates and Sandstones
	Pre-Cambrian	No definite classification yet made	Sandstones, Slates and Volcanic Rocks

standing at high angles to the horizontal. Rocks affected in this way are known as *slates.*

If we could flatten out all the beds of England, and arrange them one over the other and bore a shaft through them, we should see them on the sides of the shaft, the newest appearing at the top and the oldest at the bottom, as in the annexed table. Such a shaft would have a depth of between 10,000 and 20,000 feet. The strata beds are divided into three great groups called Primary or Palaeozoic, Secondary or Mesozoic, and Tertiary or Cainozoic, and the lowest of the Primary rocks are the oldest rocks of Britain, and form as it were the foundation stones on which the other rocks rest. These are spoken of as the Pre-Cambrian rocks. The three great groups are divided into minor divisions known as systems. The names of these systems are arranged in order in the table, and the general characters of the rocks of each system are also stated.

With these preliminary remarks we may now proceed to a brief account of the geology of the county.

The geology of Herefordshire is commonly spoken of as remarkable for its simplicity from the fact that it is nearly all upon an Old Red Sandstone formation, which gives the same ruddy tone to its surface as prevails in south and mid-Devon. In its history, however, it differs from the "Devonian" red sandstone in a striking and interesting manner; for, unlike the latter, the Herefordshire rock has never been under the sea, but formed the bed of freshwater, or possibly brackish lakes.

In the fossilised remains, therefore, found in the "Old

Red" lacustrine formation of Hereford are the freshwater equivalents in flora and fauna of those found in the otherwise similar formation in Devonshire and in parts of Scotland and Ireland. The Old Red Sandstone of this class extends almost throughout South Wales, and covers by far the greater part of Herefordshire, extending from the scarp of the Black Mountains on the south-west to the borders of the Malvern range on the north-east, and again from Tenbury on the north to Goodrich on the borders of Dean Forest, where it meets the Carboniferous limestone of that coal-bearing region. In the matter of antiquity the Old Red Sandstone comes between the older Silurian and the later Carboniferous limestone. It lies above the former on the east and the west of the county, while in the south it runs under the coal-bearing limestone of the Forest of Dean. The rocks of this formation attain a considerable thickness, particularly in the central part of the county, where it is estimated at 6000 feet. They consist of red and yellow sandstones which are quarried at Cradley, Weston, Lugwardine, and elsewhere, red marls, calcareous sandstones and limestones known as "cornstones," and beds of quartz conglomerate. The fossilised plants of the Old Red Sandstone are of peculiar interest, for the latter contains the earliest specimens of vegetation on the earth's surface—marsh plants and ferns—the ancestors of those species which attained a dominant position in the flora of the succeeding Carboniferous age. The fauna is represented chiefly by fish remains, of which some have been found in the lower beds of the series at Leominster. The other fossils are chiefly remains of crustacea and freshwater molluscs, nearly allied to the existing freshwater mussel. The most prevalent type of the "Old Red" of which the county so largely consists is that known as the "Cornstone" formation, for the reason that it is interlaid with bands of earthy limestone, called cornstones. To the resistance which these have afforded to the wear and tear of denuding agencies we owe the many and beautiful eminences which are a leading feature of Herefordshire scenery, such as Dinedor, Dinmore, Credenhill, Ladylift, and others.

Some rocks of a much greater age than those of the Old Red Sandstone protrude in places. At the Herefordshire Beacon on the Malvern Hills, the earlier gneiss appears to have been thrust over Silurian rocks which extend to Ledbury, while further south Cambrian shales and quartzites intervene. Silurian rocks also enter the county at May Hill on the south-east and again to the north at Ludlow, Aymestrey, and Leintwardine. These localities are famous hunting grounds for the typical fossils of this formation. The outcrops of limestone in these areas are remarkably fossiliferous. The valley of Woolhope (which has given its name to the well-known naturalists' club of Herefordshire, that has for a long time done very good work in many departments and published a great deal of valuable matter) is the most interesting bit of Herefordshire to the geologist, in whose language it used to be termed a "valley of elevation." Here what is called an anticline of Silurian rocks protrudes through the overlying red sandstone. This is in fact at the bottom of a valley, overlooked by escarpments of limestone which almost encircle it. The upcast of the upper Silurian rocks in this Woolhope valley forms a kind of dome about two miles in length and rather less in width, known as Haugh Wood. The south-east corner of the county trenches on the hard carboniferous limestone of the Forest of Dean, which is accountable for the deep gorges and sculptured cliffs beneath which the Wye cuts it way, and also for the rugged grandeur of the scenery. The Malvern Hills, whose western edges lie in Herefordshire, are an upheaved mass of crystalline granitoid and gneissic rock, and are regarded as undoubtedly among the oldest formations of our island.

7. Natural History.

It will be well before touching upon this subject to recall the fact that existing submarine remains and geological science have virtually proved that at no great distance of time, as geology counts time, the British islands were part of the Continent of Europe. At some later period again it is equally certain that our country was entirely submerged, and when the waters again fell it had consequently to be stocked anew with animal and plant life. This supply was naturally drawn from the unsubmerged mainland to the south and east. But, later, the Channel and the North Sea joined, and the period of union with the Continent was of no great length as time is counted in such matters—not long enough at least for our islands to get as much of the flora and fauna from such sources as they might in a longer period have been capable of receiving. Under the same law of redistribution it is natural that those portions of our country nearest to the Continent should be richer both in animal and vegetable life than those more distant from it; the south for instance than the north, and more particularly the island of Britain than that of Ireland, though neither acquired the number of species existent in France or Germany. But all this must be understood as applying to times long before history.

And, coming to quite modern times, it must be remembered that the habits, customs, and laws of different nations, particularly those relating to land, have greatly influenced both plant and animal life. The conditions of Great Britain and Ireland have hitherto been especially favourable to the preservation of such life in many of its forms. The private ownership of land in large estates and its cultivation in large or comparatively large farms, together with almost universal game preservation, have made of this country one vast refuge for all kinds of birds save a few species that are injurious to game. Besides the immense area of woodland and coppice rarely trodden from year's end to year's end, there are in Great Britain and Ireland what no other country in the world possesses, namely thousands of miles of hedgerows that, one might almost say, are held in full and virtually unmolested possession by birds and wild flowers. These, it may be remarked, are of comparatively modern growth. A hundred years ago there was probably not a third, two hundred years ago not a tenth, of the hedgerows we now see. In the Middle Ages there were none to speak of. Till we have seen other countries it is impossible to realise what a difference this makes. All these things combined make Great Britain, and particularly England and Wales, a paradise for birds and wild flowers. As great a variety of both may be found in other countries of Europe, but seldom such profusion of each species.

The character of a country, of course, to a great degree determines the character of its bird, animal, and plant life. In regard to birds, Herefordshire, as a warm rich region bordering on high cold altitudes, offers attractions to most of our native birds, except of course those which never leave the sea coast. The leading feature of the county is its abundant woodland, much of which was originally primitive forest, and we consequently find woodland birds abundant. The birds peculiar to our mountainous or semi-mountainous counties are also found here—the curlew, the golden plover, and the ring ousel are all denizens of the fringes of the Black Mountains, where grouse and black game also occur, though very sparsely within the limits of the county.

As all the streams in the county are "moorland born" and run quickly over stony beds, the water ousel or dipper is seen on them all the year round, while the sandpiper and yellow wagtail haunt them in the breeding season. All the common song birds abound in the county, blackbirds, thrushes, the finches, linnets, and buntings. Wood-pigeons and starlings have become so numerous as to be almost a pest. Of the rarer birds such as the siskin, the pied flycatcher, and the grass-hopper-warbler, each is inclined to frequent certain parts of the county in fair numbers. Both the barn and the tawny owl are common. The long-eared and short-eared owl are much scarcer, though reinforced by migrants at certain periods. Three or four pairs of ravens usually breed in the Black Mountains within the county limits, and of the scarcer hawks, the buzzard, the honey buzzard, the peregrine falcon, and the merlin are all occasionally seen. The kite and the hen harrier are practically extinct.

The county can claim the distinction that at Graftonbury, near Hereford, the British yellow-necked wood-mouse was first discovered by Mr de Winton.

The wild cat and marten are extinct and the polecat probably nearly so, and the hare has become comparatively scarce. Of foxes, stoats, weasels, polecats, and squirrels, Herefordshire has its normal share, and more than its share of otters and badgers, but it has one possession that is a rarity in England. For in the densely afforested range known as Bingewood Chase and High Vinnals, behind Ludlow, the fallow deer run quite wild, though there are no aboriginal deer.

Herefordshire is famous for its freshwater fisheries. The merits of the Wye as a salmon river are too renowned to need description, while there is probably no river in England that contains such a variety of good fish—salmon, grayling, trout, pike, perch, and chub. The Monnow, Lugg, and Arrow are all noted trout and grayling streams, and the best portions of the Teme are

Coracles on the Wye

within the county. An interesting link between the ancient and modern fisherman is the coracle—the light, oval-shaped boat, made of canvas stretched on a framework of laths, and carried on the fisherman's shoulders. Julius Caesar has left a description of this type of boat as he found it in Britain, and it may still be seen in use on the Wye, the Severn, and other Welsh rivers.

In the generous soil and forcing climate of Herefordshire it is only natural that the flora should be lavish. The richest localities in the county will be found in the woodlands and their borders, especially the shady banks below woods. Next to these will be the banks of the rivers and those of the numerous streamlets which join them. Within the county too are happily found those sharp contrasts for botanical study afforded by the prevailing red sandstone of the lower districts, as opposed to the mountain valleys of the western districts and the stupendous cliff gorges of the Wye upon the south. In the western districts about the Golden Valley the high banks of lane or brook, almost buried in summer time with masses of fern and foxgloves, form a characteristic feature. In the region to the south of the Malverns the unusual spectacle may sometimes be seen of an entire field, abandoned at some former day by the farmer, ablaze from end to end with wild roses. The shallows of the Wye are turned in summer into veritable gardens of white water-crowfoot (*Ranunculus fluitans*). Among rarities, the snowdrop, the monkshood, and the yellow star of Bethlehem (*Gagea lutea*) all occur wild, and a very rare orchid *Epipogium aphyllum*, found in the county, has been only twice met with in England. As elsewhere in the west country where the clear mountain streams come through the budding woods in spring, their moist margin is resplendent with primroses, violets, and celandine, while banks of wild hyacinth are only waiting to follow them with their blaze of blue. Later the blue cranesbill, the wood vetch (*Vicia sylvatica*) and the liquorice

Seven Sisters Rocks, on the Wye

vetch (*Astragalus glycyphyllus*) will delight the eye of the lover of wild flowers in mid-Herefordshire. Wild lilies of the valley are to be found near Symond's Yat. The blue iris flourishes upon the Cornstone banks, and in brambles Herefordshire is peculiarly rich, claiming more species than any county in England. Mistletoe is extremely common, and upon the Black Mountains, besides their glorious mantles of heather, the common whortleberry flourishes in great abundance, with the pink cowberry (*Vaccinium vitis-idaea*) and the crowberry (*Empetrum nigrum*). Besides the common ferns there are many species rare in most other counties; among them, the sweet mountain (*Aspidium oreopteris*), the oak and beech and limestone ferns, and others of rarer occurrence.

Herefordshire is essentially a wooded county, and every forest tree reaches here the perfection of maturity. But the oaks and elms are perhaps the great trees of the Welsh border counties. There are two native oaks and five native elms; and two native lime trees, one of which is accounted a great rarity. The tangled woods which clothe the lofty and almost precipitous cliffs of the lower Wye in those matchless reaches which begin at Symond's Yat, the fantastic shapes and abounding variety of foliage present a spectacle that has no precise counterpart anywhere else in England, made yet more striking by the sombre yews that here and there dot the vast bright-coloured canopy which autumn spreads upon these wondrous steeps.

8. Climate and Rainfall.

The climate of a country or district is, briefly, the average weather of that country or district, and it depends upon various factors, all mutually interacting; upon the latitude, the temperature, the direction and strength of the winds, the rainfall, the character of the soil, and the proximity of the district to the sea.

The differences in the climates of the world depend mainly upon latitude, but a scarcely less important factor is proximity to the sea. Along any great climatic zone there will be found variations in proportion to this proximity, the extremes being "continental" climates in the centres of continents far from the oceans, and "insular" climates in small tracts surrounded by sea. Continental climates show great differences in seasonal temperatures, the winters tending to be unusually cold and the summers unusually warm, while the climate of insular tracts is characterised by equableness and also by greater dampness. Great Britain possesses, by reason of its position, a temperate insular climate, but its average annual temperature is much higher than could be expected from its latitude. The prevalent south-westerly winds cause a movement of the surface-waters of the Atlantic towards our shores, and this warm-water current, which we know as the Gulf Stream, is one of the chief causes of the mildness of our winters.

Most of our weather comes to us from the Atlantic. It would be impossible here within the limits of a short chapter to discuss fully the causes which affect or control weather changes. It must suffice to say that the conditions are in the main either cyclonic or anticyclonic, which terms may be best explained, perhaps, by comparing the air currents to a stream of water. In a stream a chain of eddies may often be seen fringing the more steadily-moving central water. Regarding the general north-easterly-moving air from the Atlantic as such a stream, a chain of eddies may be developed in a belt parallel with its general direction. This belt of eddies or cyclones, as they are termed, tends to shift its position, sometimes passing over our islands, sometimes to the north or south of them, and it is to this shifting that most of our weather changes are due. Cyclonic conditions are associated with a greater or less amount of atmospheric disturbance; anticyclonic with calms.

The prevalent Atlantic winds largely affect our island in another way, namely in its rainfall. The air, heavily laden with moisture from its passage over the ocean, meets with elevated land-tracts directly it reaches our shores—the moorland of Devon and Cornwall, the Welsh mountains, or the fells of Cumberland and Westmorland—and blowing up the rising land-surface, parts with this moisture as rain. To how great an extent this occurs is best seen by reference to the map of the annual rainfall of England on the opposite page, where it will at once be noticed that the heaviest fall is in the west, and that it decreases with remarkable regularity until the least fall is reached on our eastern shores. Thus in 1908, the maximum rainfall for the year occurred at Llyn Llydaw in the Snowdon district, where 237 inches of rain fell; and the lowest was at Bourn in Lincolnshire, with a record of about 15 inches. These western highlands, therefore, may not inaptly be compared to an umbrella, sheltering the country farther eastward from the rain.

The above causes, then, are those mainly concerned in influencing the weather, but there are other and more local factors which often greatly affect the climate of a place, such, for example, as configuration, position, and

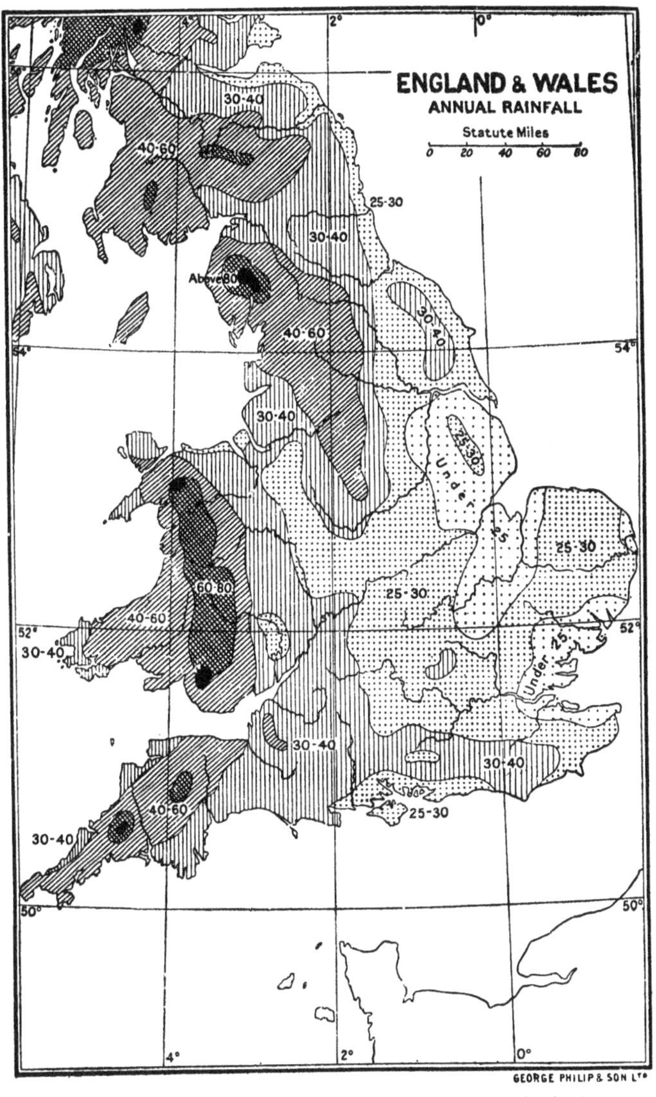

(*The figures give the approximate annual rainfall in inches.*)

soil. The shelter of a range of hills, a southern aspect, a sandy soil, will thus produce conditions which may differ greatly from those of a place—perhaps at no great distance—situated on a wind-swept northern slope with a cold clay soil.

The climate of Herefordshire is remarkably equable. But, though the county ranks in mildness of atmosphere with the other counties of the south-west, its average rainfall is far less than that of its immediate neighbours to the west or again than that of Devon and Cornwall. The Black Mountains and Radnor Forest, which form a practically continuous wall along its western edge, seem to act as a protection against an excessive rainfall. And with regard to these uplands it is noteworthy that even on their loftiest points overhanging Herefordshire, such for instance as Waun-fach, 2650 feet above sea-level, the measurements over a term of years, excluding the universally abnormal season of 1910, averages only about 53 inches, a figure which is nearly or quite doubled in great numbers of comparatively low-lying inhabited valleys in the Lake Country and Snowdonia. It is considerably less indeed than that recorded in many populous valleys in the adjoining counties of Brecon and Glamorgan, some of which average from 80 to 90 inches.

The highest readings of the very moderate rainfall of Herefordshire are, as one would expect, on its western fringes—Kington and neighbourhood under Radnor Forest with from 33 to 36 inches is just perceptibly the wettest district. The Golden Valley on the other hand, near the foot of the Black Mountains, only shows an average of

32 inches for the last four years, excluding the abnormal year 1910 when nearly 14 inches of rain fell during November and December in some parts of Herefordshire and elsewhere in England.

At Kentchurch again, at the south-west extremity of the county, beneath lofty local hills, and near the Black Mountains, the rainfall is but little over 31 inches. Going eastward, there is a great drop to the figures of Hereford and its neighbourhood which, taken collectively, give an average of about 23 inches. This central part of the county is much the driest, as is shown in the accompanying map of the rainfall. For going northward to the Leominster district, which borders on Shropshire, we find a fairly general average of 26 inches. The western division towards the Malvern Hills shows in its northern portion, represented by Bromyard and neighbourhood, a slight increase upon this, while at Ledbury again, and in the country round, the rainfall is approximately that of the Leominster district.

Though more liable to cold spells than south-west Wales or the south-western peninsula of England, the winters of Herefordshire are generally mild, while in periods of extreme heat the readings of Herefordshire are seldom or never among those that attract exceptional notice. Of sunshine there are no available records for the county.

9. People—Race, Dialect, and Population.

We have no written record of the history of our land carrying us beyond the Roman invasion in B.C. 55, but we know that Man inhabited it for ages before this date. The art of writing being then unknown, the people of those days could leave us no account of their lives and occupations, and hence we term these times the Prehistoric period. But other things besides books can tell a story, and there has survived from their time a vast quantity of objects (which are daily being revealed by the plough of the farmer or the spade of the antiquary), such as the weapons and domestic implements they used, the huts and tombs and monuments they built, and the bones of the animals they lived on, which enable us to get a fairly accurate idea of the life of those days.

So infinitely remote are the times in which the earliest forerunners of our race flourished, that scientists have not ventured to date either their advent or how long each division in which they have arranged them lasted. It must therefore be understood that these divisions or Ages—of which we are now going to speak—have been adopted for convenience sake rather than with any aim at accuracy.

The periods have been named from the material of which the weapons and implements were at that time fashioned—the Palaeolithic or Old Stone Age; the Neolithic or Later Stone Age; the Bronze Age; and the Iron Age. But just as we find stone axes in use at the present day among savage tribes in remote islands, so it must be remembered the weapons of one material were often in use in the next Age, or possibly even in a later one; that the Ages, in short, overlapped.

Let us now examine these periods more closely. First, the Palaeolithic or Old Stone Age. Man was now in his most primitive condition. He probably did not till the land or cultivate any kind of plant or keep any domestic animals. He lived on wild plants and roots and such wild animals as he could kill, the reindeer being then abundant in this country. He was largely a cave-dweller and probably used skins exclusively for clothing. He erected no monuments to his dead and built no huts. He could, however, shape flint implements with very great dexterity, though he had as yet not learnt either to grind or polish them. There is still some difference of opinion among authorities, but most agree that, though this may not have been the case in other countries, there was in our own land a vast gap of time between the people of this and the succeeding period. Palaeolithic man, who inhabited either scantily or not at all the parts north of England and made his chief home in the more southern districts, disappeared altogether from the country, which was later re-peopled by Neolithic man.

Neolithic man was in every way in a much more advanced state of civilisation than his precursor. He tilled the land, bred stock, wore garments, built huts, made rude pottery, and erected remarkable monuments. He had, nevertheless, not yet discovered the use of the metals, and his implements and weapons were still made of stone or bone, though the former were often beautifully shaped and polished.

Between the Later Stone Age and the Bronze Age there was no gap, the one merging imperceptibly into the other. The discovery of the method of smelting the ores of copper and tin, and of mixing them, was doubtless a slow affair, and the bronze weapons must have been ages in supplanting those of stone, for lack of intercommunication at that time presented enormous difficulties to the spread of knowledge. Bronze Age man, in addition to fashioning beautiful weapons and implements, made good pottery, and buried his dead in circular barrows.

In due course of time man learnt how to smelt the ores of iron, and the Age of Bronze passed slowly into the Iron Age, which brings us into the period of written history, for the Romans found the inhabitants of Britain using implements of iron.

We may now pause for a moment to consider who these people were who inhabited our land in these far-off ages. Of Palaeolithic man we can say nothing. His successors, the people of the Later Stone Age, are believed to have been largely of Iberian stock; people, that is, from south-western Europe, who brought with them their knowledge of such primitive arts and crafts as were then discovered. How long they remained in undisturbed possession of our land we do not know, but they were later conquered or driven westward by a very different race of Celtic origin—the Goidels or Gaels, a tall, light-haired people, workers in bronze, whose descendants and language are to be found to-day in many parts of

Scotland, Ireland, and the Isle of Man. Another Celtic people poured into the country about the fourth century B.C.—the Brythons or Britons, who in turn dispossessed the Gael, at all events so far as England and Wales are concerned. The Brythons were the first users of iron in our country.

The Romans, who first reached our shores in B.C. 55, held the land till about A.D. 410; but in spite of the length of their domination they do not seem to have left much mark on the people. After their departure, treading close on their heels, came the Saxons, Jutes, and Angles. But with these and with the incursions of the Danes and Irish we have left the uncertain region of the Prehistoric Age for the surer ground of History.

At the Roman invasion of Britain Herefordshire, in common with much of South Wales, was occupied by that brave people the Silures, who gave more trouble and offered a more prolonged resistance to the invaders, than almost any other people in the island. Like the rest of the population—of South Britain at any rate—they were doubtless a mixture of the earliest inhabitants of the island of whom we have any trace, the Iberians, as they are usually designated, the small dark men with long skulls, who used stone implements only, and their Celtic conquerors. These were larger, fairer men with round heads, enjoying the superiority of bronze weapons, and both the great branches of their race invaded the Silurian territory—the Goidels (or Gaels) and the Brythons (or Britons). After twelve or fourteen centuries of intermingling and relationship towards one another of which we know nothing, the Iberian stock, it is thought, was more prevalent among the Silurians than among the races further east, when the Romans came. But this is all surmise. For general purposes we now class all the races of Ancient Britain together and usually speak of them as Celts, or when dealing, as here, with Wales or the borders of Wales—as Welsh. The chief question of race in Herefordshire, which as a border county is in this respect exceptionally interesting, lies between Welsh and Saxon, though there was no doubt a certain amount of foreign blood left by the Roman soldiers of various nationalities who for three centuries were in great strength as a garrison at Caerleon in Monmouthshire, and scattered in small posts elsewhere, as in the no doubt Romano-British cities of Kenchester near Hereford, and Uriconium in Shropshire.

But the real race interest arises out of the Saxon invasion and the long conflict between the Saxons of Mercia and the Britons or Welsh who then inhabited Herefordshire. No one really knows to what extent the Saxon invaders of Britain expelled, slaughtered, or assimilated into their own communities the native British. But the opinion is generally accepted that the nearer we approach Wales in middle or southern England the stronger strain there will be of Celtic blood. Herefordshire in Saxon times became the scene of constant war. The retiring Britons there reached the natural strongholds of the country, the Welsh mountains, and uniting with the local Silurian Britons, who clinging always to these hills were probably hardier than the

Romanised Britons of the English plains, stood permanently at bay. More than this, they were able to make things very unpleasant for the Mercian Saxons occupying the larger and eastern half of Herefordshire. It was then, about 780, that Offa, the most famous of Mercian kings, cut the dyke, which bears his name, from north to south through the county, passing about four miles to the westward of Hereford town, to Bridge Sollers, after which the Wye was probably the boundary. This for a long time divided the Welsh from the English, though among the latter there were probably quite as many Welshmen who as serfs, or by choice as freemen, remained under English rule. The western or actually Welsh part of the county fell in time under Saxon rule, but merely by a transfer of allegiance to Saxon Earls and without any hardship or uprooting. Thus modern Herefordshire undoubtedly contains a very large proportion of Welsh blood. During the civil war in the seventeenth century Parliamentarian officers relate that there was as much Welsh as English spoken in the streets of Hereford.

But we have stronger evidence than all this of a pure Welsh population who remained permanently in Herefordshire. If the reader will look at the map of the county, he will note what a great number of Welsh place names there are to the south of Hereford, such as Llandinabo, Pencoyd, Hentland, Kilpeck, Pontrilas, Llangarren, and such like. In contemporary documents, as late as Glyndwr's wars in the early fifteenth century, frequent allusions are made to "Archenfield" or

"Irchenfield" as a military community to be reckoned with or attacked. Domesday book tells us about these people and their particular territory, which like the county has more or less the shape of a diamond. Its two western sides are formed by the Worm Brook and the Monnow, its eastern sides by the twisting Wye. Its northern point almost touches Hereford, while its lower extremity actually touches Monmouth. This was the district of Archenfield.

Now when the Mercians drove the Welsh over Offa's Dyke, the Silurian Welsh of Archenfield were permitted to remain unmolested, with the privilege of retaining their laws, their customs, and their language. They had probably encouraged this treatment in some way, and thoroughly justified it by a staunch loyalty to their Saxon ruler, earl or king, and a most determined hostility towards their Welsh brethren beyond the Dyke. They remained as it were a little state within a state, under their own Welsh laws, but sending six of their number as representatives to the Shire Mote at Hereford. The most curious privilege of those cherished by them is thus expressed, "When the Army marches against the Enemy (the Welsh) they form by custom the vanguard in the advance and the rearguard in the return." The Welsh of the 40 odd parishes of Archenfield, then, were never disturbed, nor were those of the adjoining district of Ewyas, afterwards attached, which came into the county under much the same curious conditions.

The greater part of the proper names in south Herefordshire and a considerable proportion of those throughout

HEREFSCIRE.

TERRA REGIS. In [illegible]

Rex. W. ten [illegible]. Rex Edw tenuit. Ibi erant. v. hidæ. 7 reddeb quartā partē firme unius noctis. Modo ē ualde imminutā. Ibi sunt in dnio. iii. car. 7 x. uill 7 v. bord cū. xii. car. Ibi. vi. serui. 7 molin de. viii. denar.

Ibi. i. francig ten dimid hidā que reddeb. iii. sol T.R.E.

Hoc ā sicut ē modo redd. x. lib de albis denar.

De hoc ā ten S. maria de cormeliis ecclām 7 presbim cū sua tra. 7 totā decimā. 7 un uillm cū una virg terræ.

De ipso ā ten Ansfrid de cormelis. ii. hid. 7 ix. uill 7 ix. car.

7 Wills. f. Baderon ten unā virg træ. que ibi iacuit T.R.E.

Ilbert uicecom ht ad firmā suā de Arcenefeld cōsuetudines omis mellis 7 ouiū. que huic ā pertineb T.R.E.

Witts. f. normanni ht inde. vi. sextaria mellis. 7 vi. oues cū agnis. 7 xii. denarios.

In [illegible]

Rex ten [illegible]. Rex. E. tenuit. Ibi. iiii. hidæ. In dnio sunt. iii. car. 7 ix. uill 7 iii. bord 7 un seruiens regis. Int oms hnt. x. car. Ibi. iii. serui. 7 molin de. x. solid.

Hoc ā redd. x. lib m de alb denar T.R.E. n fuit posit ad firmā. 7 ideo ignorant quantū ualuit. / De cormeliis. Decimā huj ā 7 un uillm cū. i. uirg træ ht S. maria.

Una ex his. iiii. hid fuit 7 est in Reuelande. Ibi sunt. iiii. bord 7 una ancilla cū. ii. car. 7 ibi. ii. molini de. xv. solid.

De aliis. iii. hid posuit Radulf de bernai. L. acras ad suū Reuelande. 7 un bord. 7 molin de. vii. solid.

Qđ ē habet uicecom de suis opib: ualt. lx. solid.

Rex ten [illegible]. Rex. E. tenuit. Ibi. iiii. hidæ. In dnio sunt. ii. hidæ una v minus. 7 ibi ē una car. 7 alia posset. ēē. 7 vi. uilli cū pposito 7 iii. bord 7 un fab. Int oms hnt. vi. car. Ibi Silua nōe Triuelde. nulla reddens csuetudinē nisi uenationē. Villi T.R.E. ibi manentes. portab uenationē ad hereford. nec aliud seruitiū faciēb. ut scira dicit.

Huj ā decimā ten S. maria de cormeliis 7 un uillm cū. i. uirg terræ. Ilbert f. turoldi ten de hoc ā. ii. hidas. p uno ā. Huic ā pertineb T.R.E. una pars træ [illegible] 7 csuetudo de eā paceb in Chingestone. Roger de Laci ten de rege.

Ad hoc ā apposuit uicecom tpr. W. comitis [illegible]. Hoc ā tenuit Aluuin. 7 potuit ire ad quē dnm uolebat.

[illegible]

Rex ten [illegible]. In dnio sunt. v. [illegible] bord cū. xvii. [illegible] Ibi. ii. molini de. [illegible] De silua 7 pastura [illegible] [illegible] De hoc ā ten [illegible] de. ii. hid. 7 Rog[illegible] de. i. hida [illegible] Ilt Rog ten dim[illegible] hanc trā [illegible] De eod ā ten [illegible] [illegible] T.R.E. ualb. vi. [illegible]

Rex ten Lene. [illegible] In dnio sunt [illegible] cū. xvi. car. Ib[illegible] Ibi. ii. molini [illegible] Villi huj [illegible] iii. sext frumti [illegible] 7 un obolū. [illegible] de cormeliis in [illegible] T.R.E. ualb. vi[illegible] pposit huj ā [illegible] in ā [illegible] 7 dapifer 7 ali[illegible]

Rex ten [illegible] Ibi. xvii. [illegible] In dnio sunt. ii[illegible] hi uilli arant [illegible] frumti 7 [illegible] ad Willm. f. b[illegible] In hoc ā ē p[illegible] iii. car. Ibi [illegible] Ibi molin nil [illegible] Ibi Silua [illegible] solid. Ad here[illegible] xviii. sextar[illegible] De hoc ā est [illegible]

Portion of Domesday Book relating to Herefordshire

the rest of the county are Welsh. One can only conclude, therefore, that the people of Herefordshire have at least a very strong admixture of Welsh blood in their veins, deadly enemies though they were in ancient times of the Welsh people beyond the Dyke, and English as they stoutly maintain themselves to be to-day with all the fervour or race pride that remains among rival border races long after they have made up their quarrels.

The Norsemen probably made no lodgement to speak of in Herefordshire, but the Normans occupied the county as conquerors with much more completeness than they did most other English counties. Being comparatively few in number it is not generally considered that they introduced much actual new blood into the Anglo-Celtic-Danish stock of the island, greatly as they influenced its character. Neither did the Flemish weavers nor the Huguenot refugees reach Herefordshire in any appreciable numbers, nor bring into it such a strain of fresh blood as they introduced into the adjoining counties of Gloucestershire and Wiltshire, or into Sussex, Kent, and Norfolk. The present population of the county may therefore for all practical purposes be set down as a blend of the Mercian Saxons and Silurian Welshmen—though we must not forget that during the last half century there has been a greater tendency to move about than in all the preceding centuries.

The dialect of Herefordshire is a distinctly pleasing one to any unbiased ear reasonably familiar with all the chief dialects of England. It belongs to the family of speech that may be called South Saxon, which with variations extends from Sussex inclusive to the borders of South Wales, taking in the counties of Hampshire, Berkshire, Wiltshire, Dorset, Gloucestershire, fringes of Oxfordshire, Worcestershire, and parts of Somerset, till it meets the English-speaking Welsh. This is a rough but sufficiently accurate definition of a type of dialect, upon which, however, the cockney accent of London in parts of the home counties has made terrible encroachments. We should all pray that it may never reach Herefordshire, the danger being that within the huge London radius it is conspicuously caught and perpetuated by quite well-educated people. Parts of Somerset and Devon speak a kindred tongue to what I have called the South Saxon, but with such marked differences of accent and method of utterance that it should undoubtedly be classed apart. When we touch the midlands again we find another kind of speech and intonation, which keeps on changing as we proceed north or north-westward till we meet dialects that a Herefordshire man would at first scarcely understand. The South Saxon speech offers by comparison no difficulties to the stranger. In those portions of which Wiltshire is the centre, the old people still have a South Saxon form of speech sufficiently broad to puzzle an outsider. But when we cross the Cotswold Hills, though it is the same type of speech, it is much less pronounced. And about Cheltenham, Gloucester, and Evesham, we get the first touch which bespeaks Welsh blood or contact with people of Welsh blood. This shows itself in a sudden spring to a higher note on the last word of the sentence. When we get into Herefordshire and more particularly as we advance into the county, we find a marked Welsh intonation or sing-song running all through it, and this is natural when the history of Herefordshire is remembered. The South Saxon speech is soft and slightly drawling, with a soft burr of the "r," a broad "a," and an "s" more or less inclining to a "z." This Saxon speech with a Welsh intonation is the distinguishing mark of Herefordshire in varying degrees, as it is of West Shropshire, Monmouthshire, and even the fringe of English-speaking South Wales. The Welsh sing-song as it sounds to a Saxon ear is derived from the correct accentuation and intonation of the here long-forgotten language of Wales, transferred to another tongue that has no affinity with it.

Some words of purely Welsh origin occur in the Herefordshire dialect. Such are "gwethal" or "gweddil" equivalent to the Italian *roba*, and meaning things or stuff; "pant" (a hollow); "pill" (a creek); "prill" (a streamlet); "mawn" (peat); "tump" (a hillock, oonty tump = a mole-hill); "ross" (a bog); "suck" (= *sweh*, a ploughshare) and others.

10. Agriculture.

There are in Herefordshire altogether 538,924 acres. Of this area in 1911, 127,205 acres were in crop or rotation grasses, and 321,105 in permanent pasture. The remaining 90,614 acres are woodland, orchard, mountain, heath, or ground occupied by buildings, roads, gardens, or water. To consider first the county's production in cereals, we find that in 1911 there were 21,242 acres in wheat, with an average yield to the acre for the last ten years of 30½ bushels, a trifle below the average of England, which is 31 bushels over the same period. In wheat production Hereford stands 31st among the 41 English counties and 23rd in yield per acre, Lincolnshire, the premier wheat county, growing 775,894 quarters, with an average of 34 bushels to the acre.

In oats we find Herefordshire ranking 33rd with 24,641 acres, and an average yield for the ten years of 37½ bushels per acre, Cambridgeshire being first with an average of 52 bushels.

From 17,688 acres of barley 74,000 quarters are produced, averaging 32 bushels per acre, Lincoln again at one end of the scale growing nearly a million quarters and Westmorland at the other under 2000.

Of beans and peas there are nearly 6000 acres and of potatoes 1400 acres. Turnips and swedes account for 14,800 acres, and mangolds for 4500 acres. The produce from clover, sainfoin, and the grasses under rotation is cut from 27,169 acres and yields 26,000 tons of hay, while 91,700 acres of meadow give 88,000 tons. There are about 5000 acres in hops, producing 41,000 cwt.

We have left hops for the last mention, as they are one of the peculiar products of Herefordshire; this county with Worcestershire constituting the western hop district of England, as Kent and Sussex comprise the more important one of the south-east. For Kent with a yield in

1910 of 186,895 cwt., and Sussex with 22,878 cwt., show more than thrice the combined product of Herefordshire, 41,000 cwt., and Worcestershire, 28,666 cwt. But Herefordshire is only half the size of Kent, so the importance of this industry to the county is rather more than half what it is to Kent. The cultivation is more followed in the eastern than in the western half of the county, and the oast-houses there are almost as familiar an object as in a Kentish landscape. The picking season is a busy and crowded time. As Londoners swarm into Kent, so Birmingham and other midland centres send every September their thousands of pickers to the Herefordshire hop gardens. Though wild hops were used from earliest times in England for the making of beer, it does not seem that the hop was cultivated in this country till the time of Henry VIII.

The orchards of Herefordshire are of course famous, but they have been cultivated far more for cider-making than for table or cooking fruit. The county first gained its reputation in the time of Charles I, when planting went forward so vigorously, that in the words of an old writer it became "one entire orchard." It now shares the preeminence with Devonshire, and the respective merits of the cider of the two counties are a matter of rivalry, though that of Hereford is now, as two centuries ago, more popular with the outside public. With Herefordshire must be included parts of the counties of Worcester and Gloucester. The quality and produce of the Herefordshire orchards might be enhanced by greater attention to pruning and culture, but much improvement has taken place since the passing of the Technical Instruction Act of 1889. Pears too are grown largely in Herefordshire, and perry (pear cider) is a delicious drink almost peculiar to the county. Most fruits and garden stuff do well in Herefordshire in its quick and kindly soil and benignant climate. But intensive small-fruit and vegetable culture for the great markets, which has made

A Cider Orchard

the neighbouring district of the Vale of Evesham so prosperous and celebrated, has not been greatly practised in our county.

Of live stock in the county in 1911 there were 25,021 horses, Herefordshire ranking 18th in this respect among the counties of England. In cattle she comes

Hereford Bull

28th with 103,277; in sheep 16th with 358,928, and in pigs 28th with 29,894.

The culture of soft fruit is receiving a good deal of attention, and an increasing acreage is being given to strawberries, black-currants, logan-berries, etc.

The pastures of the county are rich and extensive. The Hereford breed of cattle is a very old one and is world-famous. The rich deep-red cattle marked with white, with white faces, throat, and chest, and waxen-looking horns, are familiar to every Englishman or English colonist, and to every American of either the Northern or Southern Continent who knows one ox from another. In spite of the fact that they are beef rather than dairy cattle, Herefordshire remains so loyal to its own distinctive breed that we see comparatively few of any other sort upon its pastures, whereas in most parts of England nowadays we find a very great variety. The origin of the Hereford breed is not clearly known, but from the prevalent breed of the seventeenth century it was brought to its present pattern and character by one or two well-known families of Herefordshire farmers, in the neighbourhood of King's Pyon in the eighteenth century. It is the thriftiest and hardiest of all the first-class breeds, though not so quick to put on weight as the modern shorthorn under the most favourable conditions. But in wild new countries, where difficulties or hardships have to be faced by cattle, it is the prime favourite. At home it may be said to be part of the pleasant Herefordshire landscape, and as a breed it is exceptionally free from tuberculosis.

The county has practically lost its once notorious breed of sheep, the "Ryeland," whose wool, in days when mutton was cheaper but English wool of the first importance, fetched far the highest price in England. It was so valuable that it was known as "Leominster ore," and the poet Drayton thus celebrated it—

> "Where lives the man so dull on Britain's farthest shore,
> To whom did never sound the name of Leominster ore,
> That with the silkworm's web for smallness doth compare."

Cattle Market, Ledbury

It was crossed almost out of existence with Merinos increasing its fleece on the one hand, and with the Leicester increasing its weight on the other. There are a few pure bred flocks, however, still in existence and some prospect of the breed being again revived. All sorts of sheep, as everywhere else, may now be seen in Herefordshire, but the Black-faced Shropshire Down is probably the favourite. There are no wild hill pastures, except where the county fringes on the Black Mountains. So, outside these limitations, there is no sheep farming as a special industry, though grass farming generally prevails nowadays over tillage. The grazier again is more to the front than the dairyman, and the grazing of cattle is more important than that of sheep. Herefordshire farms run to all sizes, large, moderate, and small, but 150 acres is roughly a typical holding.

11. Industries and Manufactures.

Cider and perry are the only articles of any importance that Herefordshire turns out for export, but for these the county has been celebrated since at least the seventeenth century, and probably earlier. A book published early in that century, entitled *Herefordshire Orchards, a pattern for all England*, speaks of fruit culture as long established, and of every house in the county from the greatest to that of the poorest cottager as being surrounded by orchards; with a life's experience of cider and cider-making and a knowledge of all the English varieties, the author of the book claims that the liquor of Herefordshire is the best, not only in his own opinion, but in that of "all good palates." A little later, Lord Scudamore and other gentlemen were so active in promoting the industry that John Evelyn said of the county that "in a manner it hath become one continuous orchard." Everyone, indeed, drank cider and perry from the highest to the lowest, and there is evidence that a good deal was exported. Many instances are recorded by old writers of extraordinary prices paid for these beverages. It appears that the orchards of Herefordshire were to a large extent either neglected or grubbed up during the period of abnormally high prices

Cider Works, Hereford

for grain and other farm produce obtained during the Napoleonic wars.

The cider industry declined, and efforts were vainly made to call attention to the neglected orchards and the planting of new ones by that celebrated horticulturist, Mr Knight, of Downton Castle, who in the early nineteenth century was to Herefordshire what Lord

Scudamore, of Holme Lacy, had been in the seventeenth. The merchants too, who bought the cider from the farmers, adulterated it with water and fortified it with spirit, thus bringing discredit on the local article. It seems that a true knowledge of the principles of fermentation and the science of cider-making had ceased to exist. Things have improved immensely of late years and modern science in cider-making has more than restored such arts of the earlier times as had been lost. The simple rule of absolute cleanliness has been realised, and pains are being taken to produce the best species of fruit for both cider and perry. Improved machinery which saves time and labour has also been introduced, and though in the matter of apple culture there is still room for improvement throughout the county, matters are mending greatly. There is also a decided inclination on the part of the public to look with increasing favour on cider as a wholesome and palatable drink.

In former days the liquor, which was chiefly manufactured for home use, was crushed out in crude fashion in a huge trough with a rolling stone moved by a single horse, and manipulated in a rude wooden press. The apparatus, or its remnants, may still be seen—occasionally even in operation—in many an old Herefordshire homestead. The apples in all stages, green, ripe, or rotten, were thrown in together, and small attempt at cleanliness was observed either in the manufacture or the cask. That cider dependent on these methods and still more on the varied capacities of innumerable farmers would create and keep an outside market and stimulate the demand for a wholesome drink was of course impossible. Even butter, where the skill and pride of individual homesteads may fairly be assumed to be much greater, suffers so much from the same uncertainty that butter factories in expert hands, which give a uniform and dependable product, are in all countries capturing the markets.

Cider even more requires capital, skill, and enterprise for developing its best qualities. Of late years several firms have assisted in greatly popularising the use of Herefordshire cider, especially one large factory which covers several acres. In such places as this, where apples are crushed by the thousands of tons, under the best hygienic principles, and treated with modern skill and science in vats holding as much as 50,000 gallons, a very different article from the haphazard product of the homestead is of course turned out.

Blending of apples is carefully studied in these great establishments, and for this purpose the orchards of Devon and Somerset are frequently laid under tribute by the Herefordshire makers. There is now quite a large export trade done both in cask and bottle—the apparent costliness of bottled cider or perry being due to the time, care, and risk incurred in bringing the finished article to perfection.

Cider varies greatly in character, from the rough "hard" kind that is usually popular with men engaged in manual labour, to the smoother, sweeter sorts that are more generally liked by persons of a different habit of life; while the high-class bottled cider that the county is now making a name for all over England is a beverage that few people of any kind, or either sex, could fail to appreciate. In spite of the activity in cider-making, however, there are quite a number of breweries in the county.

Malt vinegar brewing is carried on extensively, for home and export trade, at the rapidly-growing village of Colwall, where also a fruit-bottling and tinning industry is in successful working, and a branch of an important mineral water manufactory utilises the excellent water from the Hill springs.

The county, as already shown, being almost wholly an agricultural one, there are practically no other industries, except those usually found under like circumstances, such as quarrying, tanning, brick and tile making, and the like.

12. History of the County.

In a former chapter something was said of the brave resistance of the Silures—of whose territory Herefordshire was a part—to the Roman arms. Under their valiant chief Caractacus they were driven slowly backwards, as is supposed, from their fortified outposts on the Malvern Hills. The traces of this long struggle, in the middle of the first century, of which the Roman historian Tacitus gives some account, are still to be seen in a chain of fortified camps throughout the county. The final battle, in which the Roman general Ostorius overthrew Caractacus, is generally thought from the description of Tacitus to have taken place at Coxwall Knoll on the Teme near Brampton Bryan. For twenty years after this the Silures gave the Romans much trouble, but eventually Herefordshire (as we now call it) was brought under the long peace of Roman rule, extending over three centuries. During this the Romano-British city of Magna or Kenchester (near Hereford) sprang up, just as the greater city of Uriconium arose to the north of it in Shropshire. After the Romans left, about 420, the Britons of these parts, so far as we know, were not troubled by foreigners for 150 years, till the desolating invasion of the heathen Saxons destroyed Magna and Uriconium, and annexed the country between the Wye and the Severn. But this, though accompanied by pitiless slaughter and dreadful ravage, did not result in a really permanent and complete occupation. It remained for the Mercian kingdom, pushing across the centre of England in the seventh century, to include these only half-annexed districts in its dominion, and thus Hereford became a true border country.

It was now the aim of the Mercians (Mercia meaning March or Border) to subdue the Britons, now called the Welsh, of the mountains beyond. But this was not greatly persevered in, and the mountaineers in their turn became continual invaders and raiders of the neighbouring Saxon territory. The Mercian conquest or annexation of this Severn and Wye country, however, was quite a different thing from the ruthless heathen invasion of the century before. In this, heathen hatred of Christianity as well as barbarism were combined, but the Mercians were by this time Christians, and the difference in their treatment of these later conquests was immense. The Welsh or Britons, though rated even legally as inferiors, were nevertheless regarded as free men, as fellow Christians, and protected by the same laws, or, as in the case of the district of Archenfield already described, by their own laws. This explains why Herefordshire and the neighbouring districts became permanently and contentedly Saxon, though so largely occupied by Britons. It was not so with the mountainous country beyond, now called Wales, which was never really conquered, and (what was still more important) was never settled in by Saxons. It remained an alien and always hostile country to the Saxons, and even more than most nations of that day, always quarrelling within itself. Offa, King of Mercia, as we have seen, threw up his great Dyke, which runs through West Herefordshire, as a boundary, though the Welsh, as already mentioned, were later pushed a little further back to the mountains.

For three centuries, till the Norman Conquest, Herefordshire, then known as a province of Mercia with the town and castle of Hereford as its centre, became a base for the frequent border wars between the Saxons and the Welsh of Wales. The Saxon-Welsh occupying fertile land, under easy laws, had no longer any sympathy with their fellow Britons from the hills, who raided them equally with their Saxon neighbours, and so they became in time absolutely English in a political sense, though retaining in some cases, as in Archenfield and Ewyas, their habits and their language for centuries. A cathedral was built and richly endowed by Offa at Hereford, which

Site of St Ethelbert's Well, Castle Hill, Hereford

however had already been the centre of a diocese for over a hundred years. This action of the Mercian king was in expiation of his treacherous murder of Ethelbert, king of the East Anglians, while his guest at his palace of Sutton Walls near Hereford.

In the tenth century this Wye and Severn country suffered from a terrible Danish invasion, but, great as was the damage done, no permanent settlements of importance were made, as in the north and east of England, and in sea-coast places elsewhere. One of the fiercest raids on Hereford, however, was made just before the Norman Conquest by a discontented Saxon earl, Algar, who with a body of Irish and Welsh under Griffith, prince of South Wales, laid waste the country, sacked and burned the city, and destroyed the cathedral, together with all its clergy who had taken refuge within it. Twenty years before the Norman Conquest, too, Edward the Confessor, half Frenchman, half monk, granted land in Herefordshire to certain Norman favourites, who built strong castles at Ewyas and elsewhere, and behaved in tyrannical, un-English fashion to the people, till driven out by the great Saxon, Earl Godwin, who during part of his reign had much power over both Edward and the kingdom. It was he who cut off Herefordshire from Mercia as a county and earldom for his son Sweyn, to be succeeded by his other son, the great Harold, of Hastings fame, who rebuilt and strongly fortified Hereford and punished its destroyers, the Welsh, by the most successful campaign that any Saxon had ever yet made throughout that entire country.

Castle Tump

(*Where the pre-Conquest Norman castle of Ewyas Harold stood*)

This was the last notable event of the Saxon period. With the Norman advent Herefordshire fell beneath the Conqueror's sway, and was placed under the rule of his most intimate friend and supporter, Fitz-Osborn, whose advice is said to have brought about the invasion of England. Castles were built all over the shire, Norman garrisons introduced, and the county became ultimately a base for the conquest of South Wales, which proceeded by slow degrees. Indeed, this last conquest was not achieved by the king and a royal army at all, but by Norman adventurers with charters from successive kings to take such territory as they were able from the Welsh princes and hold each man his own conquest, ruling it absolutely, under fealty only to the king himself. Thus grew up to the west of Hereford what was known as the Marches of Wales, a number of petty kingdoms, as it were, governed by "lords marchers" and by laws of their own, and outside the "king's writ." Some territories were left in the hands of Welsh princes and nobles under the same condition. The Welsh, who had by comparison never been seriously annoyed in their own seats by the Saxons, fought desperately against these much more determined conquerors. Constant insurrections, or wars between the two nations, provoked by fresh aggressions, sometimes provincial, sometimes on a big scale, kept Wales in a turmoil for two centuries. It is necessary to explain this in order that the reader may understand what position Herefordshire held as an English shire always face to face with this hornet's nest, and constantly attacked by the native

Welsh till the final conquest of Wales in 1282, and then again in Glyndwr's wars of 1400 to 1410. In most English shires the country people during that period, from lack of practice and of enemies, had become quite unwarlike. But those on the Welsh borders were all, so to speak, soldiers. They were constantly fighting. They and the South Welsh were the first to bring the long bow—later the common weapon of England—into general use. They were practised and formidable archers before the rest of England, with certain exceptions, had any skill to speak of with that weapon. The power which such a command of great numbers of war-hardened and skilled men gave the great Marcher Houses, like the Lacys, the Despensers, the Clares, and the Mortimers, in addition to their influence over, and often blood alliances with, the Welsh chieftains, was of great weight in the destinies of England and the fortunes of its kings. Herefordshire then formed an important part of that fraction of England and Wales to which many successive kings in troublous times looked for their stoutest supporters or their most dangerous foes. In the Barons' Wars of Henry III, Llewelyn, prince of North Wales, laid waste West Herefordshire, and it was from Hereford that Prince Edward, placed there as a hostage for his father's agreement with the barons, made his famous escape, under the ruse of a horse race, to the Mortimers' castle at Wigmore.

The county was much concerned, too, with the hapless Edward the Second's wanderings. His hated favourite, Hugh le Despenser, was hanged at Hereford on a gallows 50 feet high, while the triumphant Mortimer, arbiter of the kingdom and favourite of Queen Isabella, kept high state at Wigmore. For the Welsh Marches after 1282, save in Glyndwr's wars (1400–10), had practically ceased to be the scene of English and Welsh conflicts as such. The South Welsh of the score or two of independent lordships formed the chief following of these Anglo-Norman lords, and were equally ready to fight against Scots, French, English, or each other, besides supplying thousands of trained mercenaries to the English kings. In Glyndwr's wars, when Wales broke out once more in almost universal flame, the men of Herefordshire received the first attack, and fell to the number of 1100 in the battle of Pilleth on the Lugg. During several years the county was frequently harried by the Welsh, while English armies often mustered in Hereford, which city continued in a state of constant alarm.

In the Wars of the Roses, the House of York being heirs of the Mortimers, and so great Herefordshire landowners, the county had an ample share in the fighting. The two great Yorkist victories, at Mortimer's Cross, near Kingsland (after which the captured Owen Tudor was beheaded at Hereford) and that of Tewkesbury, were won within 20 miles of the city. With the Tudor dynasty, whose advent gratified and entirely reconciled the Welsh, and with the transformation by Henry VIII of the powerful, semi-independent Marcher baronies of mid and South Wales into ordinary counties, the importance of Herefordshire as a military and political centre of national moment ceased. Its castles were mostly

Memorial Stone on the battle-field of Mortimer's Cross

abandoned to the bats and owls. Its people laid aside their bows and spears for good, and subsided, like their turbulent neighbours, into the peaceful, unmolested, humdrum life that the people of most other English counties, save in the intervals of national civil wars, had been leading for centuries. But the personal licence, natural to ages of fighting and semi-lawlessness, could not at once be exterminated. The borders of Herefordshire and her neighbours were infested for some time by outlaws and refugees, often members of good families, who gave constant trouble. To check this turbulence, as well as for other reasons of policy, a sort of High Court was set up in Ludlow Castle by Henry VII, known as the Court of the Marches of Wales, administered by a President and Council, who were always high dignitaries. Herefordshire and the other Border counties came under this government, which generally supervised the affairs of Wales, and continued, though with declining importance, till the time of William III.

In the wars of King and Parliament, Herefordshire was chiefly in sympathy with the former. Hereford city, however, was held for the Parliament by Lord Stamford at the opening of the war, but was soon after abandoned to the Royalists, from whom, however, Waller snatched it in 1643. Upon the Parliamentary troops being again called elsewhere, Hereford was re-occupied by the other party.

The people of the county, like those of many others, grew weary in time of the exactions made on behalf of the King's cause. When Charles, after the defeat of

Naseby, came to Hereford to revive the loyalty of the Borderers, they responded only moderately, but the city, reinforced by 2000 men, held out against the Scottish Covenanters supporting the Parliament till the Scots were called away to the north. After this it was captured by surprise by the Parliamentarians at Christmas, 1645, and this was really the last event of any historical note Hereford or the county witnessed. The great change in importance, however, came to Hereford with Henry VIII. As in all other counties, the abolition of the monasteries and the establishment of the Protestant religion was an immense upsetting, for there were many religious houses in the county. But the turning point in Herefordshire history, perhaps, was the conversion of Wales into a peaceful and really united partner of England, instead of a complicated medley of royal counties and independent ill-governed lordships. Till then every English king, from the Conqueror onwards, had been in Herefordshire, most of them many times, it being a notable post of danger and observation. From now onward, save in the special case of Charles I, only an occasional royal visit, and that merely of pleasure or compliment, can be recorded, for the old political and military importance of the county had disappeared, while its small population, and its remoteness from London, which now became more and more the centre of England, placed it on another footing from that which it occupied in the Middle Ages.

13. Antiquities—Prehistoric, Roman, Saxon.

By far the most prominent reminders of the various races who lived in Herefordshire before any written history begins, are the ramparts and ditches which formed the main part of their rude fortresses and still encircle so many hill tops. These are for convenience called British camps, though many of them, no doubt, were here in some form long before the Silures so bravely confronted the Roman legions. Who threw up the first entrenchments, whether Iberians in the Stone Age, the Silurian tribes of the later Celtic-Iberians in the Bronze and Iron Ages, or whether in a few cases the Romans or even the Saxons themselves, no one can say with any certainty. Probably most of the prominent hills now bearing "camps" were occupied by all in turn, and altered to suit the fashion or the immediate needs of each race. The earlier and more primitive people, we may be well assured, used their hill camps not merely against human tribal foes, but as folds safeguarding their stock as well as themselves, during the night, from wild beasts that prowled in the thick and pathless forests covering the valleys. Rivers did not then run as the Wye runs now, in a regular confined channel, with clean dry pastures and cornfields spreading smoothly away on either hand, but their waters, checked, turned, and dammed back by the fallen wreckage of tangled forests, oozed over great areas of wet woody marshes, keeping the primitive population more or less on the dry uplands, probably grazing their stock in the ranker herbage below during the day-time, and doubtless themselves hunting or fishing there.

In some places the trackways where for long ages cattle have moved up and down from hill settlements to the lowland have been cut into deep lanes. In times of

The Herefordshire Beacon

danger from human foes the utility of the hill tops needs no explanation. Nowadays in Herefordshire most of the isolated hills rise like high wooded islands out of a comparatively open country, but in those times they were most likely bare hills rising out of universal forest. But these camps, as ages passed, were no doubt improved for the purely military purposes of a somewhat advanced native people, and much more advanced invaders—sometimes by the Romans, often probably by the Mercian Saxons, or again by the Danes. The finest camp in Herefordshire is quite bare and is certainly British, and is one of the most nobly placed, conspicuous, and remarkable in all England. It stands on the Herefordshire Beacon, one of the lofty Malvern hills, 1114 feet above sea-level. Another notable camp is on Dinedor Hill, the existing works of which are thought to be Roman and to have been thrown up by Ostorius Scapula; while Aconbury is another one near at hand. Wall Hill is a pentagonal camp of 39 acres near Ledbury. Credenhill covers a wooded eminence seven hundred feet high and has a double ditch encircling 30 acres, just above the Romano-British city of Kenchester. The camp on Wapley Hill (1100 feet) on the Radnor border is one of the finest "elliptical" fortifications in England and is one of those almost certainly occupied by Caractacus. Coxwall Knoll, close to Brampton Bryan, where he is thought to have made his last stand as described by Tacitus, is lower, in a picturesque and wooded valley, but it has an exceptionally strong and elaborately constructed camp enclosing 30 acres. These are but a few of the numerous hill fortresses in Herefordshire, some oblong, some oval, some circular, which besides the ramparts, now of course greatly diminished from their original height, were doubtless in many instances further strengthened by stockades of post or brush. From these camps and other places, as elsewhere in England, have been gathered both flint and other stone implements, spear-heads, arrow-heads, hammers, scrapers of the Neolithic or New Stone Age, and the more elaborate implements of the later Bronze and Iron Ages. Among these last are swords and spear-heads, pottery, ornaments, coins, and trinkets. The use of bronze is thought to have commenced about 1400 B.C.

Arthur's Stone

but how long before the time of the Romans iron came into general use, can only be a matter of conjecture.

There are "barrows" or burial mounds to be seen here and there in the county, but its very luxuriance has tended to obliterate their traces under plough or woodland, more than is the case on the down land of counties such as Wiltshire, Hampshire, and Dorset. There is a famous cromlech, however, in Herefordshire, known as Arthur's or King Arthur's Stone, situated above the Golden Valley near Dorstone; a massive block of sandstone 18 feet long, resting originally on eleven smaller uprights, some of which have fallen. This, like other cromlechs in Britain, was almost certainly the grave of some famous chieftain, and, like a somewhat similar one in Glamorganshire, is associated with and named after King Arthur, the great half-legendary hero of South Wales.

There were three Roman towns or stations in the county. Ariconium at Bollitree near Ross, and Kenchester at Credenhill near Hereford, with a third, Bravinium, occupying the site of Leintwardine. Kenchester, the Magna Castra of the Romans, was much the most important and was the capital of the district. Its site, which covers 20 acres, is raised a little above the surrounding country, and being all open field is very plain to the eye, which can readily follow the outline of the vanished walls almost round the five sides that enclosed the city. In a very dry summer the marks of the foundations of the streets and buildings can be seen in places. A large number of relics have been gathered from here at different times, many of which, together with those from other parts of the county, can be seen in the museum at Hereford, and others are even yet being frequently ploughed up. These large stations were not merely Roman garrisons, but trading and residential towns, inhabited by Britons who, during three centuries of peace, learned many of the arts and luxuries of Roman life. They were sacked and destroyed by the first great invasion of the pagan and comparatively barbarous Saxons. These last-mentioned people had then no appreciation of large towns or fine buildings, living themselves in isolated communities and in rude wooden houses.

Hereford is thought to have been originally built of the stones carried from the ruined city of Kenchester. All three of the Roman stations in Herefordshire were on the great Roman road known as Watling Street which will be spoken of in a later chapter. Saxons to the last built almost wholly of wood, except in the case of some of their churches. So nothing but a few portions of their work in these remain, and they will be alluded to under the heading of Architecture. One famous reminder, however, of the Saxon period, exists in the earthworks upon the site of King Offa's Palace known as Sutton Walls, near Marden on the Lugg. Offa's Dyke, too, has been already mentioned. As an earthen rampart with a deep ditch on the Welsh side it may be traced for some miles northward, from Bridge Sollers on the Wye six miles west of Hereford, and southward near Symond's Yat. On the breast of Great Doward Hill is "Arthur's Cave," where the teeth and bones of rhinoceros and hyaena were discovered in 1871, relics of a time earlier even than the stone arrow-heads, and not to be overlooked in any mention of prehistoric remains in Herefordshire.

14. Architecture—(*a*) Ecclesiastical.

The churches of a country side should assuredly, as a whole, be the most interesting of the objects erected upon it by the hand of man. They are practically all—in origin, if not always in actual fabric—much older than the oldest dwellings. They are generally more ancient in one or both senses than even the oldest castles. But the castles are for the most part in ruins, and have long been dissevered from any human interest. Within the same church walls, on the other hand, and always at any rate upon the same spot of ground to which the people gathered centuries ago, their remote descendants or successors still gather for precisely the same purposes of worship, and upon those other vital occasions that mark the ceaseless progress of birth, marriage, and death. Here and there, to be sure, are secular buildings that recall the story of centuries, and we very naturally and rightly make much of them. But interesting as their story is, it is usually that of families or individuals. The church has few tragic or stirring tales to tell, but it has been, in a sense, the common property for all these ages of an entire community, and has remained unchanged in its ownership, neither bought nor sold, nor captured, nor confiscated. Mere changes in method of worship count for nothing against the great fact of its continuous presence as the central point of the parish. This may seem, perhaps, to have been more conspicuously so when every one was compelled, by moral or legal pressure, to attend public worship, when the churchyard was the natural place for parish gatherings and even for parish games, and still more when, as was often the case, its tower was the parish fortress against enemies. But, even in the absence of these conditions, and not forgetting the recent rise of other sects seceding from the particular form of doctrine taught within its walls, the parish church remains a building, the historic interest of which is not always fully realised from the very fact of its every day familiarity. Its presence in every parish is taken for granted. It is an almost too familiar object everywhere, and people are apt to forget that in this very familiarity, and in this quiet indifference to the ups and downs of secular life and the buildings around it, lies its charm, even apart from all religious influences. For the history of the parish for long centuries is written here. The lord of the manor, the yeoman, the tenant farmer, the tradesman, and the labourer, are all here in brass or alabaster, in marble or simple stone or wood, in effigies sometimes, in carved inscriptions by the score or hundred, within or without the walls, and still more abundantly in nameless graves. But they are all here, generations of them, while for three or four centuries, at any rate, a more complete and a less exclusive record of the main facts in the lives of both the lowly and the great lie in the parish registers. The parish church, in a sense, has a deeper meaning, and gives more to think about apart from the presence or absence of architectural interest, than even the old historic mansion, but it is of such familiar association to us that its significance is very apt to be forgotten.

The rememberance of this side of the importance of parish churches helps to promote an interest in the various styles of architecture which distinguish them. Their styles denote various periods, and anyone with a very little trouble can learn enough to tell at a glance to what period a church, or its various parts (for most of them have been frequently added to, or restored) belongs.

A preliminary word on the various styles of English architecture is necessary before we consider the churches and other important buildings of our county.

Pre-Norman or, as it is usually, though with no great certainty termed, Saxon building in England, was the work of early craftsmen with an imperfect knowledge of stone construction, who commonly used rough rubble walls, no buttresses, small semi-circular or triangular arches, and square towers with what is termed "long-and-short work" at the quoins or corners. It survives almost solely in portions of small churches.

The Norman conquest started a widespread building of massive churches and castles in the continental style called Romanesque, which in England has got the name of "Norman." They had walls of great thickness, semi-circular vaults, round-headed doors and windows, and lofty square towers.

From 1150 to 1200 the building became lighter, the arches pointed, and there was perfected the science of vaulting, by which the weight is brought upon piers and buttresses. This method of building, the "Gothic," originated from the endeavour to cover the widest and loftiest areas with the greatest economy of stone. The

Norman Nave, Leominster Church

first English Gothic, called "Early English," from about 1180 to 1250, is characterised by slender piers (commonly of marble), lofty pointed vaults, and long, narrow, lancet-headed windows. After 1250 the windows became broader, divided up, and ornamented by patterns of tracery, while in the vault the ribs were multiplied. The greatest elegance of English Gothic was reached from 1260 to 1290, at which date English sculpture was at its highest, and art in painting, coloured glass making, and general craftsmanship at its zenith.

After 1300 the structure of stone buildings began to be overlaid with ornament, the window tracery and vault ribs were of intricate patterns, the pinnacles and spires loaded with crocket and ornament. This later style is known as "Decorated," and came to an end with the Black Death, which stopped all building for a time.

With the changed conditions of life the type of building changed. With curious uniformity and quickness the style called "Perpendicular"—which is unknown abroad—developed after 1360 in all parts of England and lasted with scarcely any change up to 1520. As its name implies, it is characterised by the perpendicular arrangement of the tracery and panels on walls and in windows, and it is also distinguished by the flattened arches and the square arrangement of the mouldings over them, by the elaborate vault-traceries (especially fan-vaulting), and by the use of flat roofs and towers without spires.

The medieval styles in England ended with the dissolution of the monasteries (1530–1540), for the Reformation checked the building of churches. There

Kilpeck Church: Norman Doorway

succeeded the building of manor-houses, in which the style called "Tudor" arose—distinguished by flat-headed windows, level ceilings, and panelled rooms. The ornaments of classic style were introduced under the influences of Renaissance sculpture and distinguish the "Jacobean" style, so called after James I. About this time the pro-

Ducking Stool, Leominster Church

fessional architect arose. Hitherto, building had been entirely in the hands of the builder and the craftsman.

Having indicated these various styles, however, it must now be borne in mind that a majority of churches needed restoration or enlargement long before the Reformation. This work was done, as the earlier work had been done, by the monks of the monasteries and various orders, helped by skilled masons who travelled about and took immense pride and pains in their work. So a majority of English churches contain work belonging to different periods. The greatest church-building age was just after the Norman period, in the twelfth and thirteenth

Abbey Dore Church

centuries. The greatest restoring and re-building age was in the Perpendicular period, and in the wealthier and more fertile parts of the country particularly, this style is more conspicuous in churches than the earlier ones which it displaced. Norman, Early English, and Decorated are the prevailing styles in Herefordshire and are all exemplified in the cathedral (see p. 136). Though there is a certain amount of Perpendicular work, there is not very much building of original fifteenth century date. Of Saxon work there is scarcely any, but a good deal of Norman, and even where the original churches have been greatly altered, the round-headed doorways frequently survive.

Of the Norman style Kilpeck and Moccas are the most

Bromyard Church

complete specimens among Herefordshire churches, the former having few, if any superiors, of its class in England. It was rebuilt 60 years ago every stone being carefully replaced in its original position, even those mutilated being neither rejected nor refaced. The old divisions of choir, nave, and sanctuary are here distinctly marked and the eastern apse is one of the most perfect in the country. A striking external feature is the row of grotesque heads of men and beasts, over 70 in number, that extend the whole way round the building upon a corbel table beneath the eaves. At the west end are three brackets representing crocodiles' heads supported by their tongues.

Pembridge Church
(*showing detached belfry*)

The south door also is a beautiful specimen of Norman work and moulding. The plan of Moccas church, which has also been restored, was identical with that of Kilpeck. It contains some fine Norman arches and a sculptured tympanum of curious design.

Ledbury Church

The nave of the noble priory church of Leominster, otherwise much altered and rebuilt in various styles, is a fine example of the Norman style on a large scale. The origin of this church was a Saxon priory founded by the West Mercian Earl Leofric. The great width of the building is due to the inclusion of a later and an older building under the same roof. The church contains an interesting relic in the form of a ducking-stool, used in earlier days for the punishment of scolds and witches. The use of this and of similar instruments of punishment lasted until the beginning of the nineteenth century, and it is interesting to note that the two last recorded victims, Jenny Pipes, "a notorious scold" (1809), and Sarah Lecke (1817), both belonged to Leominster.

The most complete example upon a stately scale of the Early English style is the monastic church of Abbey Dore which is practically all that is left of an important Cistercian foundation. The building is Early English throughout, except the tower, which was added much later. The great nave—the foundations of which are clearly traceable—was possibly never completed. The existing portions of the church, which was cruciform, are transept and choir with several aisles and chapels. At the intersection of the nave and transepts are four arches of quite unusual span, while the lancet windows of the choir are of exceptional size. A remarkable feature of the building is a double and vaulted aisle running right across the east end of the church behind the choir. The choir aisles are also vaulted, as was originally the roof of both choir and transept.

Colwall Church

Bosbury (Bishops-burgh, once an episcopal seat) is a good example of Early English, with a massive detached tower, but it contains some perpendicular windows. Fownhope, on the Wye, retains a central Norman tower but is otherwise Early English. So is Peterstow, though preserving much Norman work and erected on the foundations of a Saxon church. Stoke Prior has an Early English chancel and a Norman nave.

Madley, one of the largest and finest churches in the county, is a conspicuous example of the mixture of successive styles. A portion of the nave is Norman with its cushioned, scalloped capitals. The windows are partly Early English and partly Decorated. The chancel is mainly Decorated, with a polygonal apse, and beneath it is a crypt having a groined roof supported by a central shaft. Bromyard has a cruciform church finely uplifted with an embattled central tower of Norman origin. In its nave, which is mainly of later work, are still two good Norman doorways.

The large church of Ledbury is a blend throughout of every style, more than usually intermingled. There is here, too, one of the half-dozen detached towers for which the county is noted, Early English in this case and carrying a lofty modern spire. Of these detached belfries, Pembridge has the most curious example—part stone but mainly wood, and built within of the most massive timbers. The shape is unique, and slightly suggestive of a Chinese pagoda. The spacious cruciform church itself is a good example of Early English and Decorated. The tower of Colwall church is Perpendicular but the

Ross Church

body contains both Norman, Early English, and Decorated portions. Ross church is a most imposing building, beautifully situated, and conspicuous for its spire, which rises to a height of over 200 feet and contains a fine peal of ten bells. The church itself is late Decorated and Perpendicular. Kingsland should not be omitted from the briefest notice of Herefordshire churches not only for the rare Volca chamber beside the north porch, the residence in olden times, as is supposed, of a recluse, but also for its hoary, massive, and embattled tower, its spacious nave, and its chancel, all of which are of the thirteenth century. It forms a conspicuous feature over a wide landscape.

15. Architecture—(*b*) Military.

Herefordshire as a Border county is naturally very rich in the remains of medieval castles, but is less fortunate than her neighbours in their state of preservation. Of most of them, indeed, very little is left. The Saxons did not build castles in the ordinary sense of the word. The *castellum* mentioned in the older chroniclers was more of the nature of a walled town than a castle, and resembled the *castrum* of the Romans rather than the "castle" of the Normans. These walled enclosures were called by the Anglo-Saxons *burhs*, from which is derived our modern burgh, or borough.

The first stone castles built in England by private persons on their own domains were Richard's Castle, built by Richard, son of Scrob, a Norman adventurer, and Ewyas in Herefordshire, the work of Ralph the

Sketch Map showing the Chief Castles of Wales and the Border Counties

Norman, nephew of Edward the Confessor. He was one of that band of Edward's Norman favourites already spoken of who were settled in England before the Conquest, and all (except he, who became Earl of Hereford) expelled by the indignant English. They had made their strong castles centres of tyranny, a system quite alien to Saxon custom, but destined to become only too familiar; for after the Conquest the Normans built castles everywhere on the lands conceded to them under the feudal system.

The passion for building castles—"dens of the devil and nests of thieves," the old chronicler Matthew of Paris calls them—grew apace. They were convenient engines of tyranny and exaction to their Norman lords, and for keeping what they had acquired against the greed of their equally powerful neighbours and fellow-subjects. In Stephen's reign alone over 1100 are said to have been erected. But in Herefordshire most of the castles were erected as defences against the Welsh.

The normal type of the earlier Norman castles was a moated mound with a timber palisade, and its base court or bailey also moated and palisaded. This is called the motte-and-bailey type of castle, the central mound or *motte* being often artificial, though where possible a natural hill or rock was adapted to the required shape. Gradually the wooden palisading on the mound was replaced by a wall of stone, forming what is called a shell keep. Sometimes, instead of the shell keep, a massive rectangular tower was placed on the mound, as first at Rochester, and in our own neighbourhood at Goodrich and Ludlow. In the reign of Henry II both square towers and shell keeps began to go out of fashion, cylindrical towers taking their place.

Towards the end of the reign of Henry III an entirely new style of castle-building came into fashion—the elaborate "concentric" system called by the name of Edward I, which he brought to the highest pitch of perfection in the

Goodrich Castle

castles of Caernarvon and Beaumaris. We have no example of this later style in Herefordshire.

Goodrich, on the Wye, is the best survival among the Herefordshire ruins. A portion at least of every feature of a great Norman castle may here be seen. The keep is a massive square tower of impregnable strength, as at Ludlow, and is thought by some, but erroneously, to be pre-Norman. The Clares were among its owners. In the Civil War it was the last castle in Herefordshire that held for the King, after which, like many others still surviving at that time, it was unroofed and "slighted" by order. Among other castles of which portions remain

Wilton Castle

are Wigmore, the great stronghold of the Mortimers; Wilton near Ross; Eardisley; Bollitree near Ross; Kinnersley; Old Castle on the Monnow; Clifford on the Wye near Hay, whence came Fair Rosamond, favourite of Henry II; Richard's Castle; Ewyas Harold (only the mound and works of which remain); Huntington near

Kington; and Brampton Bryan, which was bravely defended by Lady Harley, in her husband's absence, against a royalist force in the Civil War. The still considerable ruins of the important border castles of Grosmont, Whitecastle, and Skenfrith, which were royal fortresses to guard Herefordshire against the Welsh, still stand, but on or near what is now the Monmouthshire bank of the Monnow.

Bollitree Castle

Many of these Herefordshire castles were held as fortresses by constables under the Earl of Hereford, representing the King in the turbulent pre-Tudor times against the Welsh, with an eye also on the powerful Lord Marchers in Monmouth, Glamorgan, and Brecon.

The earlier kings disliked the baronial castles of

England generally, as constant centres of intrigue and rebellion, but were obliged to encourage those of the Lord Marchers that protected the kingdom at their own expense against the native Welsh chieftains, though in the earlier years of the fifteenth century these indomitable fighters were at the very gates of Hereford, and kept the city in a constant state of alarm, while under Glyndwr himself in the same decade they got as far as Worcester.

16. Architecture—(*c*) Domestic.

During the sixteenth century a great change came over England and Wales. With the advent of a Welshman on the male side to the throne of England Wales had become thoroughly reconciled to the Union. The ancient prophecies of its seers were all fulfilled, and its long wounded pride was more than satisfied. The Civil War of the Roses had, moreover, killed off half the turbulent nobility of the kingdom. The Tudor monarchs, Henry VIII and Elizabeth, set themselves steadily to crush or check the power of the remnant and to encourage a new aristocracy with other ideals and a personal loyalty to the Throne, assisted by profuse grants of the confiscated church lands at the Reformation. Feudalism gradually ceased to exist. Quarrelsome Barons with few ideas but those of war and self-aggrandisement, supported by small armies of retainers intrenched in uncomfortable, impregnable castles, ceased to be an element, much less the chief element, in English life. Civilisation in short leaped forward, literature and the arts became the mode. In the far north, on the Scottish border, savage warfare still went on both between the two nations and the Borderers, but this was too remote to affect life in the rest of England and Wales. The castles were abandoned as useless, or in some cases remodelled on

Old house, Hereford

the stately and cheerful lines of domestic Tudor architecture. Wealth advanced rapidly with peace and civilisation, and new country houses adapted to the advance in manners sprang up in every direction. Scarcely any of the Herefordshire castles were remodelled, as was the case with a few of the more splendid Marcher fortresses in the neighbouring counties of South Wales. Many of the former had been already gutted and few were worth re-construction. But new houses sprang up all over the county in the reign of Elizabeth and the earlier Stuarts.

Architecture depends, of course, on the materials to hand. The Herefordshire landowners for instance built

Bargates, Leominster

their manor houses to some extent of limestone, but quite as often of timber and plastered wattle, generally known as the "black and white" or "half-timbered" style, which is more prevalent in Herefordshire than in almost any other county. The farmhouses and cottages were mainly constructed thus till quite recent times, the roof being usually of stone flags or straw. Speaking first of the greater houses, the style of that day known as Tudor, or more precisely as Elizabethan, is distinguished by its numerous gables, its projecting mullion windows and its clusters of tall slender chimneys, while the larger houses were often built round three or even four sides of a court yard. This was the real old English style. In the Stuart period it was modified by certain changes introduced from abroad into that known as Jacobean. At the beginning of the eighteenth century the "Queen Anne" style came in, as easily recognised at a glance in its widely different form as the Elizabethan. It is generally represented by a plain oblong two or three storied house with straight rows of sash windows rather longer in proportion to their width than the ordinary window, covered by a single high-pitched roof without gables frequently containing a row of dormer windows, and projecting conspicuously at the eaves, while very broad massive chimney stacks took the place of the graceful clustered shafts of former days. The Queen Anne house, large or small, with its admirable proportions, has a chaste and dignified look, and is nearly always of stone or brick. In Herefordshire houses of this type are chiefly familiar in old-fashioned country towns, representing the original mansions of prosperous citizens, or as large farmhouses, sometimes the former abode of country gentlemen. All through these periods, however, princely mansions on continental (mostly Italian) models were erected by the wealthier nobles here and there in England.

Of the many ancient country houses and granges of Herefordshire, Holme, or Holm Lacy, is the largest and perhaps the stateliest. Built of stone and brick, mainly in the seventeenth century and in the form of the letter H, it presents three noble fronts and contains an entrance hall 90 feet long. Inherited by the younger branch of the Scudamores from a de Lacy heiress early in the fifteenth century it passed from the male line in 1716 but remained till its quite recent sale in the hands of various heirs by blood.

Kentchurch Court on the Monnow, still in possession of the elder but untitled branch of the Scudamores, shares with Monnington on the Wye the distinction of being the last refuge of Owen Glyndwr, two of whose daughters married the respective owners of either seat. Kentchurch, though rebuilt on the site of an older house, is a spacious castellated, Gothic building, containing a tower of partly old work which tradition associates with the Welsh hero. The deer park climbing Garway Hill is an ancient chase, still in a primitive state and containing some immense and venerable yew trees. Monnington has probably more right to the Glyndwr tradition, and a rough flat stone in the churchyard a stone's throw from the Court is by pious repute the hero's resting place. The Court, now a farmhouse, still retains a fine oak roof, and several of the original rooms, in one of which is an elaborately sculptured fire-place. A wide avenue of ancient firs extending out into the neighbouring country is held to mark the route of Glyndwr to Wales when hard pressed by his enemies in his interludes of retirement.

There is no more beautiful specimen of a sixteenth century half-timbered house in the county than The Ley, near Weobley, a village till lately unrivalled for its half-timber work but now sadly spoiled. The Ley however has no history worth recording and has been for long a farmhouse. Hampton Court on the other hand, a large mansion near Leominster, is full of interest, though of the original house little more than the fine entrance tower remains. It was built by Rowland Leinthall, a man of only moderate condition, but as a favourite of Henry IV married a lady connected with that king and thus acquired lands and consideration. He gained yet more of both in the next reign by the ransom of many prisoners of rank taken at Agincourt, and the house at Hampton, like many others in England, was built by French money thus acquired. The house was ahead of the times in construction, having a large cistern contrived in the roof, fed by a stream conducted from the hill above. The Coningsbys, who founded the still extant almshouses at Hereford adjoining the ruins of the Blackfriars Monastery, owned Hampton in the sixteenth and seventeenth centuries. One of them, a prominent supporter of William III, was with him when wounded at the Boyne, and staunched the blood with a handkerchief still preserved by the Earls of Essex, his descendants.

Brinsop Court, owned in former days and for many generations by the Danseys, is in its way unequalled in the county, and some further human interest is attached to it from the fact that the poet Wordsworth, whose wife's near relatives were in his day its occupants, frequently visited there, and furthermore planted a cedar, now a fine tree. The Court is a large rambling moated house, part Tudor and part much earlier, now occupied by a large farmer. Its chief interest is the original fourteenth century portion in two blocks, one of which is the old banqueting hall with Gothic windows and a beautiful oak roof covered with flagstone. Treago, near

Blackfriars Monastery, Hereford

St Weonards, is also an excellent specimen of the fourteenth century manor house, owned moreover by the same family since the time of Edward II. It forms a square; the old hall, now a kitchen, retains the original timbers and arches, while there are some curious narrow doorways, secret passages, and hiding holes. Gillow manor house not far distant, now a farmhouse, is of the same date; still retaining the entrance tower and several other portions of the original building.

The old vicarage at Eardisland, continuously inhabited, is one of the most noted examples of fourteenth century half-timbered work in the county. The massive interior beams and hewn out arches which for six centuries have borne the heavy flagstone roof remain intact. Orleton Court, near Orleton village, which is rich in half-timbered work, is one of the best examples of the style remaining and dates from the sixteenth century. It was the property till recently of the Blount family—one of whom, Thomas Blount, was an industrious antiquary in the seventeenth century and left abundant and invaluable records concerning Herefordshire history which are now preserved at Belmont near Hereford. The picturesque projecting window over the porch marks the room traditionally occupied by the poet Pope when paying attention to a daughter of the house. Wythall, between Ross and Goodrich, is another fine example of the same style and period, containing in addition a remarkable spiral oak staircase, and the house in this case still remains with the descendants of its builders. Goodrich Court, though not a century old, should not be overlooked from the fact that it was built by the famous antiquary Sir S. R. Meyrick to hold his matchless collection of armour and other treasures, now dispersed among museums and collectors. It is in massive castellated style with drum towers and drawbridge, and occupies an imposing position above the Wye.

What remains of Hergest Court near Kington, now

a farmhouse, is of early fifteenth century work, but is chiefly interesting as the home at that period of Ellen Vaughan or Ellen "Gethin" (the terrible), the daughter of the then owner. This young and comely woman, so runs the tale, being goaded to vengeance at the killing of her favourite brother by his cousin John Vaughan of Tretower in Breconshire, disguised herself as a man and crossed into that county to compete at an archery meeting where "Shon hir" (Long John) was to be a conspicuous performer. While standing by him as an unrecognised stranger and competitor she bent her bow as if at the target but swinging suddenly round discharged the arrow into her cousin's body and in the confusion escaped. There was rivalry between the two branches of the Vaughans and, once safe at Hergest, Ellen became a heroine in that district. She nevertheless afterwards married a Vaughan from Tretower, who fell in the Wars of the Roses. She lies buried in Kington church and her full length effigy, notable for its decoration, may be seen lying beside that of her husband on an altar tomb in the Vaughan chapel attached to the same building.

Stoke-Edith, in the Ledbury district, built about 1700 by the Foleys, who still occupy it, is an imposing quadrangular building of red brick, while in Ledbury itself, occupying the rather unusual situation for a territorial manor house of fronting two main streets, is a spacious and beautiful half-timbered seventeenth century residence, for two centuries the home of the Biddulph family. Within is a quadrangular court, and in the park are some of the finest elms in England.

There are a few good examples of the Queen Anne period in the county. Much Marcle Rectory is one, while a still more interesting and even larger house, of rather earlier date but practically of the same style, is that utilised till recently as the vicarage at Eye. Here

Market House, Ledbury

are some magnificently decorated Italian ceilings of late seventeenth century date.

As regards village architecture, nearly everything in the county under this head that is old and picturesque is in the half-timbered style, and of this Herefordshire, from its purely agricultural character, retains even more than the neighbouring shires which share with it this feature.

The villages of Weobley, Kingsland, Pembridge, and Eardisland, lying almost in a group as they do, may be cited as exhibiting as good a display in a single neighbourhood as could be seen in the county. Several old

Market House, Ross

market halls exist. Ledbury (half-timbered sixteenth century), and Ross (of stone and Jacobean in style) are the best examples, but there is an interesting small open market-house wholly of timber at Pembridge.

The county is rich, too, in dovecotes. A Gothic half-timbered one at the Butts near King's Pyon is the most beautiful. A circular one of stone with flag roof at Richard's Castle, and a large brick one with gables at Eardisland, are two others worthy of note.

Eardisland Dovecote

17. Communications—Past and Present. Roads and Railways.

It is no exaggeration to say that modern English people, unless they have lived in some quite new country or colony, where conditions in this one respect are often the same as they were in old England, cannot realise what English roads were like before the days of McAdam

—as late, that is to say, as the beginning of the nineteenth century. It would be more to the point perhaps to say that they cannot realise the extraordinary and daily inconvenience that our forefathers suffered in this respect. Formerly, the roads of England were mere dirt-tracks, levelled occasionally and for the moment by ploughs or spades, and filled in here and there with brush or big stones. The soil through which they ran alone varied their degrees of abomination, clay being the worst and limestone or light sand the best natural beds. Travellers were literally at the mercy of the soil and the weather. In dry summers the deep ruts and hoof-marks got partially filled up by traffic, and afforded a comparatively firm though rough surface. But in winter, and in wet weather at any season, carriages, coaches, and waggons alike laboured through slush and mud, often sticking fast till local assistance and fresh horses came to haul them out. The average pace for stage coaches or light vehicles in winter, was three miles an hour, and in summer about five. The difficulties of transporting merchandise or produce were tremendous for counties without water carriage. An abundant harvest in Lincolnshire, for instance, was of no use to people in Herefordshire or Shropshire if their crops failed or were ruined by a wet ingathering. When canals came in towards the end of the eighteenth century they caused almost as much excitement and speculation as the railroads, which soon afterwards to a great extent eclipsed them. It thus came about that men and women rode on horseback for preference, and over the rough roads of those days the saddle-horses had

8—2

to pick their way and were trained to certain paces, varieties of a running walk of about five miles an hour, generally alluded to as "ambling." Pack-horses were used largely for transporting produce and travelled in long strings carrying bells. The old Roman roads were much better than any English highways between their days and those of George III. Doubtless in portions, even after centuries of neglect, they still afforded some relief to the traveller in the time of the Georges, and even later. This difficulty in travelling and conveying produce had an enormous influence on English life. It kept people at home, and to an extent now inconceivable made their district or their county the horizon by which their lives and opportunities were bounded.

The Romans made many roads through Herefordshire, of which Watling Street, one of their four main highways in England, is the most important. Coming from what in Roman times was the very considerable city of Uriconium in Shropshire, and passing through Church Stretton, it enters the county near Leintwardine and passes by Brandon Camp, the Roman station of Bravinium. Thence it proceeds by Wigmore, Aymestrey, and Mortimer's Cross, and subsequently by King's Pyon and Burghill, where turning westward it reaches Kenchester (Magna Castra). Several branches diverge from this central station. One going eastward, through Holmer, crosses the Lugg by the present bridge and thence, sometimes following the modern Worcester road, it passes through the parishes of Stretton Grandison, Castle Frome, and Cradley, after which it probably crossed the Malvern Hills into

Worcestershire. The track of these Roman roads or "strata" is marked with places which derive their modern names from their position upon it, names such as the various Strettons, Stretford, Street, New Street, and so forth. A branch of the road leaves Stretton Grandison—where there was a camp, probably Circutio—and proceeds south-east through Ashperton, where is a Roman villa, through Pixley, Aylton, and Little Marcle into Gloucestershire at Preston. From Stretton Grandison to Ashperton and beyond, this last road is now utilised as a public highway, and its singularly straight course is very noticeable when travelling along it for several miles.

Another Roman road leaves Kenchester for the south-west, crossing the Wye near Eaton Bishop and thence by Madley, where as Stone Street it is made use of to-day. After this it runs down to Abbey Dore and crosses the high ridges to the Upper Monnow valley and Longtown, where there was another camp. Hence it runs south and out of the county to Abergavenny, the Gobannium of the Romans. Yet another road enters the county from Newent on the south-east, and running just south of Ross by Ariconium, crosses the Wye near Walford, and going by Goodrich and Ganarew, again leaves the county. Many Saxon relics have been gathered at Ariconium, and at Peterstow, not far off, there are evidences of iron having been smelted. These Roman roads were very narrow, and the vehicles were built to suit them. They were carefully and elaborately constructed with thick layers of broken and unbroken stone, often consolidated with cement.

Of the principal modern roads in Herefordshire which radiate from Hereford, three run south-west, south, and south-east, to Abergavenny, to Monmouth, and through Ross to Gloucester respectively, while another follows an eastward course to Ledbury. The most travelled route of all is that which runs north, over Dinmore

Wilton Bridge, near Ross

Hill to Leominster, and thence by Wooferton junction to Ludlow.

Another important road is that proceeding due west from Hereford, more or less following the Wye into Wales, and leaving the county not far from Whitney. But the county is well supplied in every quarter by roads, the best of which are among the best in England, owing to the propinquity of the famous Dhu stone, quarried in the Clee hills, near Ludlow.

There are now no canals in Herefordshire. The railroads, as regards ownership and running powers, are somewhat complicated. The Hereford to Shrewsbury line is owned by the Great Western and London and North Western companies, each of which runs trains over it. Branches from this at Leominster, to Kington and Bromyard respectively, are worked by the G.W.R. The line from Hereford to Worcester, and that from Hereford to Abergavenny, belong to the G.W.R., but the Midland have running powers over the former, and the L. & N.W. over the latter. The Hereford to Hay line belongs to the Midland company, whose trains reach it by running from Worcester over the G.W.R. line. The Golden Valley line, from Pontrilas to Hay, is worked by the last-named company.

18. Administration and Divisions—Ancient and Modern.

County government is a very ancient institution in England and is of Saxon origin, though greatly encroached upon by the feudal system introduced by the Normans. Each county then had its Shire Moot (which met once or twice a year) and was divided into Hundreds, on the original assumed basis of a hundred families; the Hundreds, of which there are eleven in the county, being divided again into townships or parishes. Each of these smaller divisions had their courts, and even the townships had their assemblies, where every free man could appear and vote for the making of local laws; reeves and constables being appointed for the enforcement of them.

The Welsh of Archenfield in Herefordshire, for instance, though in their special case living under their own Welsh laws, sent, as we have seen, representatives to the Shire Moot of Hereford. So the County, District, and Parish Councils that have been instituted in England within recent years were by no means the innovations that people were at first apt to consider them.

Under the present mode of county government the chief officers are the Lord-Lieutenant and High Sheriff. The former is usually a nobleman or rich land-owner, while the latter is chosen every year, on the morrow of St Martin's Day, from among the land-owning gentry of the county. The County Council now conducts the main business of the shire, and at Hereford, as in other capitals, meets in its own large building, which, in addition to the County Hall, is provided with offices for all the departments under its supervision. There are 68 members of the Council in Herefordshire, of whom 51 are councillors and 17 aldermen, elected by the Council. The Council has the management of the roads, the education, and the police of the county, and exercises a general supervision over matters of sanitation. The Lunatic Asylum is jointly administered by the County and City Councils. The immediate management of certain matters, such as roads, falls to the local councils. Of these the city of Hereford and borough of Leominster have Town Councils, while Bromyard, Kington, Ledbury, and Ross have Urban District Councils. Bredwardine, Bromyard, Dore, Hereford, Kington, Ledbury, Leominster, Ross, Weobley, Whitchurch, and Wigmore have Rural District Councils. In addition to these, every parish with over 300 inhabitants, if not otherwise represented, has a Parish Council which deals with minor matters connected with it.

The law is administered in four Courts, that of Petty Sessions by the Magistrates, who refer more serious cases to Quarter Sessions, also presided over by the principal magistrate of the county. At Quarter Sessions a "Grand Jury" is empanelled, which investigates these cases in a separate room, and decides whether a "true bill" shall be found, or in other words, whether or not the case is genuine enough to send for trial by a Common Jury. Cases too grave for the magistrates, who are not of course professional lawyers, unless they happen accidentally to be such, are sent up to the Assizes. These last are held thrice a year in Hereford, and presided over by a judge, when juries are empanelled. There is also a County Court, presided over by a judge of lesser rank. This court decides cases unconcerned with crime, such as debts or broken agreements.

Petty Sessions are held at various places all over the county. Quarter Sessions are held at Hereford, while that city and the borough of Leominster have police courts of their own. County Courts are held at Bromyard, Hereford, Kington, Ledbury, Leominster, and Ross.

For the purposes of relieving the aged poor and regulating the assistance given out of the rates to the necessitous of all kinds who claim it, Herefordshire is divided into eight Poor Law Unions. Each of these is under a Board of Guardians, whose duty it is to supervise the workhouses, and appoint the officers necessary to their management and to the general working of the system.

The county is well served in educational matters.

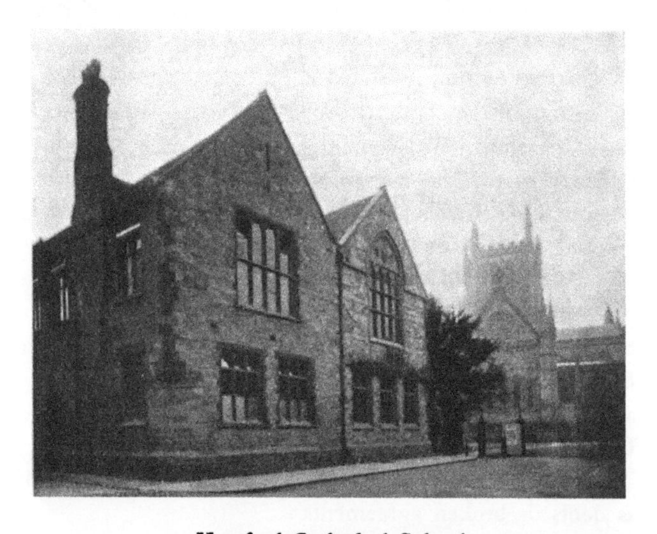

Hereford Cathedral School

In secondary education, the chief place is occupied by the Cathedral School, which is well endowed, possesses scholarships to Oxford and Cambridge, and prepares boys for the universities, as well as for professional and commercial walks of life. There are also endowed Grammar Schools at Bromyard, Kington, and Lucton. Secondary schools for boys and girls have been opened at Leominster and Ross by the County Council, and one for boys at Hereford, while one for girls is also proposed at the latter place.

There are nearly 200 elementary schools in the city and county, all but those of the former being under the control of the Education Committee of the County Council. There is a Training College at Hereford for female school-teachers, and a Blue Coat school in the town, also another at Ross, called the Walter Scott School, after its founder.

There are two Parliamentary Divisions, North and South Herefordshire, each sending one member to the House of Commons, in addition to which the city of Hereford elects a representative of its own. In ecclesiastical matters the whole of Herefordshire, most of Shropshire south of the Severn, twenty-one parishes in Worcestershire, and a few in the counties of Radnor and Montgomery, are in the diocese of Hereford, the bishopric of which dates from 676 A.D. There is an archdeacon of Hereford; and the county is divided for ecclesiastical purposes into eleven rural deaneries, which include the whole of its 245 parishes.

19. Roll of Honour.

At first sight it might appear that the birthplace of distinguished people must always be a matter of the purest accident, that talent has no connection with topography, and that, given two districts of equal area, there is no reason why one should produce a higher percentage of genius than the other. To a great extent this is the case, and indeed it is true that geography plays a more evident part in almost all the sections of which we have treated than in this. But there are instances, nevertheless, in which position and environment have not been without their effect. A peculiar maritime situation combined with excellent natural harbours made Devon the home of a race of hardy sailors and adventurers to whom honour and glory came easily, while many, if not most of our great inventors—Watt, Stephenson, Arkwright, and a host of others—have owed their position and fame to the fact of having been born in great industrial centres.

Hereford has no seaboard and no great industrial cities, and the men of distinction she has produced, as we might expect, mostly attained their fame by other paths than those afforded by such factors. A Border county however, and the theatre of ceaseless fighting, she has given birth to many warriors of more or less note, especially among the great baronial families—the Mortimers, the Laceys, the Devereux and others. Conspicuous among these, though he played many parts, was Robert Devereux, Earl of Essex, the favourite of Queen Elizabeth, whose remarkable career of adventure and plot came to a tragic end on the scaffold in 1601.

It is, however, in less strenuous and dangerous fields that the majority of notable Herefordshire men have won their fame. Taking agriculture first, the two brothers, Richard and Thomas Knight, who acquired considerable

Nell Gwynne

renown as horticulturists and planters on their estate of Downton Castle, flourished in the late eighteenth and early nineteenth century, and Sir Uvedale Price, who lived and died at Foxley, was no less distinguished as the designer and layer-out of Kew Gardens and the author of a work on the Picturesque. The praises of cider were warmly sung, in the manner of Virgil's *Georgics*, by the poet John Philips (1676–1709), who though an Oxfordshire man by birth was for some time resident in the county; and among other poets connected with it were John Davies (1565–1618), celebrated as one of the best penmen in England, and Elizabeth Barrett Browning, who came to Ledbury as a child and lived there some years before her marriage with Robert Browning the poet in 1846.

Few counties can claim more theatrical notabilities than Hereford. Here was born Nell Gwynne, "pretty witty Nell," of whom Pepys speaks so often in his immortal diary, one of the earliest of actresses, for women did not appear on the English stage much before 1660. Mrs Siddons, the greatest tragic actress of her own or any other time, was born at Brecon, but she spent all her early life in Hereford. Here, near Ross, lived Kitty Clive, the famous comic actress, and here George Stephen Kemble, brother of the more distinguished Charles Kemble and of Mrs Siddons, first saw the light. Last, but greatest of all, comes David Garrick. Born in 1717 at Hereford, the son of a captain in the army quartered there, he was educated at Lichfield grammar-school and later under Samuel Johnson. He attempted the law only to become a wine-merchant, and eventually found his true calling on the stage, where he reigned supreme for nearly 40 years, dying in 1779. Garrick was equally master of tragedy and comedy, and was worthy of his resting-place in Westminster Abbey.

David Garrick

Dick Whittington was no mythical person, whatever his cat may have been, but a real Herefordshire worthy, who was born at Sollers Hope, met with some, at least, of the happenings traditionally attributed to him, and was Lord Mayor in the years 1397, 1406, and 1419. At his death in 1423 he left money for many public objects—the repair of St Bartholomew's Hospital, the rebuilding of Newgate, and the founding of the Guildhall library, and in addition he helped to rebuild the nave of Westminster Abbey. A contemporary of Whittington was Lord Cobham, "the good Lord Cobham," better known as Sir John Oldcastle, the ardent Wycliffite and the first of our titled martyrs, who was long and intimately connected with the county, if not a native of it.

In the domain of art and letters we do not find the county very abundantly represented. John Grandison, Bishop of Exeter, who built a great part of that cathedral, was born in Herefordshire, and his name still lives in that of the parish of Stretton Grandison, near Ledbury. Another celebrity of much later date was John Abel, who, though scarcely known outside Herefordshire, nevertheless built the most beautiful timbered houses in the Border country. David Cox, one of England's greatest painters, excelling in watercolour, was for nearly 13 years resident in Hereford (1814–1827), where, primarily a teacher in a ladies' school, he gave lessons in his art to many persons, and successively occupied various houses in the outskirts of the city. During his residence here J. M. Ince, who also attained fame as a watercolour painter and was probably himself a Herefordshire man, having been born at or near Presteign on the Radnor border, became his pupil and friend. At Goodrich for many years resided Joshua Cristall

(1767–1847), more than once President of the Royal Institute of Watercolours, and one of the most dis-

David Cox

tinguished of the early school. His tomb, and that of his wife, are in Goodrich churchyard. Doubt has been cast on the authorship of the well-known heraldic work known as Gwyllim's *Display of Heraldry*, some authorities holding it to be the product of the researches of one John Barkham, but in any case Gwyllim, who was born in Hereford, was its publisher and accepted author.

Pope, in one of his *Moral Essays*, has immortalised the good deeds of "the Man of Ross." This was John Kyrle (1637–1724), a philanthropist who spent most of his life at Ross and most of his fortune in the building of churches and hospitals. Not less of a benefactor to mankind, though in a different way, was one with whom we may fittingly bring this list of county worthies to a conclusion. This was Richard Hakluyt, who came of an old Herefordshire family settled at Yatton, near Ross, and was born about 1552. He was educated at Westminster and Christ Church, Oxford, and took orders. It was to geography, however, that his life's work was devoted, and he may well be termed the father of that science. Friend of Drake, Sir Philip Sidney, Howard of Effingham, Humphrey Gilbert, Walsingham, and the leading spirits of that age, he for years patiently collected and recorded the narratives of travellers, and in 1598 began to publish them under the title *Principall Navigations, Voyages, Traffiques, and Discoveries of the English Nation*. After his death many of his unpublished papers were brought out by Purchas in his *Pilgrimes*. The Hakluyt Society, which was formed for the publication or reissue of early voyages, was founded and named in his honour. He may be regarded as one of Herefordshire's greatest men.

20. THE CHIEF TOWNS AND VILLAGES OF HEREFORDSHIRE.

(The figures in brackets after each name give the population in 1911, and those at the end of each section are references to the pages in the text.)

Abbey Dore (470), a parish near the mouth of the Golden Valley eleven miles south-west of Hereford, noted for its abbey church mentioned earlier in this book. The abbey was founded by Robert son of Ewyas in 1147, for a colony of Cistercian monks who under several enterprising abbots throve exceedingly. Though scarcely anything is left of the monastery itself, Abbey Dore is interesting for the satires flung at its monks for their supposed self-indulgence and land-grabbing propensities by Giraldus Cambrensis and Walter Mapes. It stands on the Dore. (pp. 94, 117.)

Bosbury (Bishopsbury, 852) is an old seat of the bishops of Hereford four miles north-west of Ledbury. The church is distinguished for one of the many detached towers that are a feature of Herefordshire. The late well-known novelist Edna Lyall (Ada Ellen Bayly) is buried here. (p. 96.)

Brampton Bryan (256), on the Shropshire border and the upper waters of the Teme, possesses the remains of the castle of the Harleys, one of the few Border families that in the Civil War supported the Parliament. Under Lady Brilliana Harley, in her husband's absence, the castle withstood a siege by the royalists, who after a month of unavailing effort were beaten off and obliged to retire, though the village and church were burnt. The valiant lady died soon afterwards from the strain. Harley, the first Earl of Oxford, was born in the house. A horse and pony fair is held annually in the village. (pp. 11, 23, 66, 79, 102.)

Bredwardine (247), a village beautifully situated on the Wye, seven and a half miles north-east of Hay, which is here crossed by a picturesque stone bridge. The fir-crowned hill known as the Knapp is just above, with the cromlech called Arthur's Stone near by. An old castle formerly stood here, the property of the Vaughans. There is a well-preserved effigy in the old church of one of them, Sir Roger Vaughan, who, it is said, fell at Agincourt in defending the life of Henry V. (p. 121.)

Brinsop (135), a parish five miles and a half north-west of Hereford, situated among the beautifully wooded Ladylift and Foxley hills. It is noted for its association with the poet Wordsworth, to whose memory there is a window in the ancient little church; and also for the famous moated manor house of Brinsop Court, occupied for a long time by the Hutchinsons, relatives of the poet's wife. (p. 108.)

Bromyard (1703), a small market-town on the little river Frome, twelve and a half miles north-east of Hereford, in the north-east corner of the county, without any striking characteristics except a cruciform church of Norman construction, which retains the tower and some other portions of the original work. (pp. 18, 24, 43, 96, 119, 120, 121, 122.)

Clifford (747), a village two and a half miles north-east of Hay, close to which, upon the banks of Wye, are situated the ruins of the old Norman castle associated with the Cliffords, of whom "Fair Rosamond" was one. (p. 101.)

Colwall (2010), a populous and growing residential place beneath the western slope of the Malvern hills, three and a half miles north-east of Ledbury. Immediately above it is the Herefordshire Beacon, conspicuously crowned with its very perfect prehistoric earthwork, supposed to have been occupied by Caractacus in his defence of Silurian Britain. The railway tunnel under the Malvern hills starts from Colwall parish. There is a good Early English church here, and the Herefordshire home of Elizabeth Barrett Browning was also in this parish but is now included in that of Wellington Heath. (p. 96.)

Credenhill (272), a pretty village four miles north-west of Hereford with many half-timbered houses, a church with Norman nave and some old painted windows portraying Thomas à Becket and Bishop Cantelupe. It is noteworthy for the camp on the summit of the lofty wooded hill above the village, and as adjoining the site of the Romano-British town of Kenchester or Magna Castra, already mentioned in these pages. (pp. 30, 79, 81.)

Eardisland (508), a small village on the Arrow four and a half miles west of Leominster, noted for its picturesque black-and-white houses—the old vicarage, a fourteenth century building, being unsurpassed of its kind in the county. There is also a curious and conspicuous red brick Tudor dovecote. (pp. 23, 110, 113, 114.)

Eardisley (746), a large village fourteen miles west-north-west of Hereford. It possesses a fine church of twelfth and fourteenth century work. (p. 101.)

Ewyas Harold (471), a straggling picturesque village on the Dulas one and a half miles north-west of Pontrilas, recalling in its name the influence of the Saxon in these parts and the three-cornered struggles between Edward the Confessor's Norman settlement in Herefordshire, the great Saxon Thegns, and the Welsh. The high mound on which one of these early Norman castles stood, probably displacing a Saxon fort, is a feature here, and this district of Ewyas was a Welsh community on the same terms as Archenfield with the people of Herefordshire. In another sense the parish is noteworthy as having been quite recently the subject of a history by Canon Bannister which throws light on this portion of the Welsh border. (pp. 67, 69, 96, 101.)

Fownhope (737), a large village close to the Wye, five and a half miles south-east of Hereford. It has a curious church with a Norman tower. The manor house and property have been for centuries associated with the Lechmere family. Near by is Capley Camp on the summit of a lofty hill above the Wye, noted both as a prehistoric fortress and for the magnificent view over the valley below. (p. 96.)

Goodrich (485), close to the Wye, nearly four miles south-south-west of Ross, contains the ruins of a famous castle and is the gateway to the beautiful scenery about Symond's Yat. The church with its ancient tower and spire is associated with the grandfather of Dean Swift, who was vicar here for most of his life, which included the Civil War period. He was a militant Royalist and suffered much loss and persecution on that account. A quaint house that he built and lived in still stands in the parish. (pp. 19, 29, 99, 100, 110, 117, 128, 129.)

Hereford (22,568). The county capital is situated on the Wye and derives its origin and consequence from the cathedral founded in 676, on a site rendered important by the fact of its being a notable passage over the river. Its history is identified with that of the county—briefly sketched in the text of this book. The castle was founded, or at least rebuilt, by Harold, son of Earl Godwin, who afterwards fell at the Battle of Hastings. The castle has vanished, but fine public grounds and gardens lifted above the river, commanding beautiful views of the Black Mountains and the distant Radnor moors, cover much of its site. The Shire Hall containing the County Council offices, and the Town Hall containing the City Council offices, are here, also a museum and an admirable Free Library, which includes a reference room.

The Cathedral School and the Hereford Secondary School provide higher education for boys, and a modern secondary school for girls is being built by the County and City Councils. Besides the cathedral there are two ancient churches carrying lofty spires, All Saints, and St Peter's. The Coningsby Hospital, erected for pensioners in 1614, on the site of a Commandery of the Knights of St John, is a beautiful old building.

The Choir: Hereford Cathedral

The cathedral, which crowns a slope above the Wye and replaced an earlier church burned by the Welsh, was commenced in 1079, and though not one of the larger group, exhibits fine specimens of Norman, Early English, Decorated, and Perpendicular work. It consists to-day of nave, choir, double transepts, and Lady chapel, with a massive central tower, a crypt, and cloisters. Within are several chantries, a fine library of rare old chained books and the renowned map of the world made by a Canon in 1313. The West Front has recently been well restored. The Chapter House was destroyed in the Civil Wars, but to the south of the cathedral, reached by a cloister, is the college of the vicars choral, a fifteenth century quadrangular building of much interest.

The Hereford Mappa Mundi

The Episcopal Palace, which adjoins the cathedral on the side of the river, is a very ancient building, including an old Norman hall with pillars of timber. The Wye, which is rapid and shallow just beneath the town, affords good boating for some distance above it. Bearing with it the waters of so many Welsh mountain streams, it is liable to great and rather sudden floods. (pp. 1, 2, 16, 19, 50, 67, 69, 73, 75, 76, 81, 103, 118, 120, 121, 122, 123, 126, 128, 130.)

Holme Lacy (263), on the Wye, four and three-quarter miles south-east of Hereford, and chiefly associated with the largest and most celebrated of the old family mansions of the county. It was erected in the seventeenth century and was recently sold by Lord Chesterfield, whose predecessors inherited it from the ennobled branch of the Scudamores, who had been seated here since the fourteenth century. The church is Norman and contains many ancient monuments. (pp. 63, 106.)

Holmer (515), two miles north of Hereford, is distinguished for the detached belfry of its ancient little church.

Huntington (180), a small village three and a half miles south-west of Kington on the Radnor border, containing an old church and the fragment of a border castle on a high tump. There is a noted pony fair held here annually. (pp. 12, 101.)

Kentchurch (307) is on the Monnow near Pontrilas and has an interesting old church associated with John-o'-Kent, a medieval character of familiar local fame, around whom a great deal of mystery has gathered. Kentchurch Court is in part very old, and has been the property of the elder branch of the Scudamores since the fourteenth century. One of its owners married the daughter of Owen Glyndwr, and Kentchurch contests the honour with Monnington of having sheltered the hero in his last years. The ruined castle of Grosmont, though across the river in Monmouthshire, looks down on Kentchurch from the ridge of Grosmont, upon which was fought a battle by Henry the Third in the barons' wars and another by Prince Henry against the forces of Glyndwr. (pp. 43, 107.)

Kilpeck (177) is seven miles south-west of Hereford and contains the mound and fragment of a Norman castle, together with one of the most perfect little Norman churches in England. The castle was last inhabited by the Pye family just prior to the Civil War, two succeeding generations of whom devoted their careers and sacrificed their great fortune to the losing causes of Charles I and James II respectively. (pp. 49, 91, 92.)

Kingsland (944), three and a half miles north-west of Leominster on the Lugg, is a long and important village containing several half-timbered houses and a very fine thirteenth century church. Just outside the village was fought in 1461 the great battle of Mortimer's Cross, where Edward Duke of York defeated the Lancastrians and marching on to London was proclaimed King as Edward the Fourth. An inscribed stone has been erected on the battlefield. (pp. 73, 98, 113.)

Kington (1819), prettily situated on the Arrow near the border of Radnorshire and adjacent to the beautiful scenery of Hergest Ridge, Hanter, and Knill, and to a very perfect section of Offa's Dyke. It is connected by a branch line with Leominster. It has a market-hall and is a good market, particularly for sheep and ponies, from its propinquity to the Welsh hills. It possesses an Elizabethan grammar school. The church is strikingly situated on an eminence, is of the Early English and Decorated periods, and its tower carries a lofty octagonal spire. Just outside the town and now used as a farmhouse, still stands a considerable portion of Hergest Court, erected in the early fifteenth century, and the seat for many generations of a famous branch of the Vaughan family. (pp. 23, 42, 110, 111, 119, 120, 121, 122.)

Ledbury (3358), situated in the east of the county near the base of the Malvern hills, is the market-town of a considerable agricultural district and possesses also some quarries of limestone and grey marble. There are many old half-timbered buildings, including the market, but the most striking feature of the town in this respect is Lord Biddulph's mansion, which fronts on

Church House, Ledbury

two streets, and is in itself three-sided. It has been the family residence for about two centuries and has a large park in the rear, the elms of which are among the finest in England. It is a rare instance of what is in fact an old country house, with landed estates attached, standing in a town.

The church is a very imposing one showing all styles from Norman to Perpendicular, but is chiefly distinguished for a "detached" tower with a spire, standing on the north side, and a large and beautiful baptistery. Ledbury town stands on the little river Leadon and probably derives its name from it. (pp. 24, 30, 43, 79, 96, 111, 113, 118, 120, 121, 126.)

Leintwardine (979), at the junction of the Teme and Clun near the northern apex of the county, is identified with the ancient Roman station of Bravinium, the ditch and vallum being still visible. There is also a fine Early English and Perpendicular church. The river Clun joins the Teme here and the trout and grayling fishing, for which the river is celebrated, is in great perfection under the supervision of a famous angling club. (pp. 23, 30, 81, 116.)

Leominster (5737) on the river Lugg is the second town of the county, deriving its name as is supposed from Leof-Minster, "the church of Leofric," that Mercian Earl who was husband of the celebrated Lady Godiva. The town was famous in the Stuart period, and before it, as the market for the finest wool in England (Ryeland), known poetically as "Leominster Ore." Worcester and Hereford both petitioned against its dangerous rivalry, and by getting its market-day altered put an end to its further aspirations. At a still earlier date it had been the best market in the county. Prior to that it had played the usual stormy part of a frontier town against the Welsh. It is now the centre of a large agricultural district. Its church is one of the largest in the county, and is associated, at least, with a Saxon monastery. Leominster, however, was given by Henry I to the great Abbey

Leominster Church

of Reading, which founded a cell here that became the richest in England. The church exhibits all the successive styles and is united to an older Norman church. The tower, partially Norman, is a very fine one; it stands at an angle, and contains a beautifully moulded and recessed doorway. The surrounding country is rich

Grange Court House, Leominster

in half-timbered houses, to some of which reference has already been made. At the Dissolution, all the property belonging to the priory went to the Crown and was retained in its hands for a long period. Queen Mary granted the first charter of incorporation in 1553. Before the Reform Bill, Leominster returned two members to Parliament. (pp. 16, 23, 29, 43, 60, 94, 108, 118, 119, 121, 123.)

Lugwardine (656), a village two and a half miles east of Hereford, containing a Norman church. (p. 29.)

Madley (723), a village five and a half miles west of Hereford, celebrated for its fine church, originally of eleventh century date and containing a crypt. (pp. 96, 117.)

Marden (768), in the Lugg valley five miles north of Hereford, has a good Early English church with some interesting brasses, but is chiefly noted as containing the old fortification of Sutton Walls, the site of the palace of Offa, King of Mercia. The church is supposed to cover the burial-place of the murdered King Ethelbert of Anglia before the conscience-smitten Offa moved his bones to lie under the spot where, a little later, the original of the present cathedral was built in memory of the canonised victim. (p. 82.)

Moccas (197), on the Wye, nearly ten miles west of Hereford, has a fine little Norman church and is the seat of the Cornewall family. (pp. 91, 92.)

Monnington (75), a parish on the Wye, on the opposite bank facing Moccas above Hereford, which only calls for notice as having the best claim to be the last refuge as well as the burial-place of Owen Glyndwr. One of his daughters married a Monnington who lived in the original of the Tudor manor house adjoining the church, which last was rebuilt in the seventeenth century. Tradition points to a rude flat stone outside the church porch as covering the grave of the Welsh hero, and associates his name with other landmarks in the parish. (p. 107.)

Orleton (584), a large village five miles south of Ludlow, noted for the number of beautiful black-and-white houses in the parish, of which the finest is Orleton Court, the old seat of the Blount family. The church (Perpendicular) contains some very curious old chests hewn out of the solid wood. A cattle fair is held here. (p. 110.)

Pembridge (995), a large village, abounding in half-timbered houses, situated on the Arrow, six and a half miles west of Leominster. It was formerly a market-town and contains a quaint old market-hall. The church is a good specimen of the

Pembridge

Decorated style and possesses a detached wooden belfry of an almost unique kind. (pp. 23, 96, 113.)

Peterchurch (565) is in the Golden Valley, eight miles north-north-west of Pontrilas, and the church has some fine Norman arches and contains upon the wall the sculptured figure of a trout with a gold chain round it. The remains of Snodhill Castle are in this parish.

Ross (4682) is seated picturesquely on the Wye and has some repute as a centre for visitors to the finest scenery on the English portions of the river. It is also an agricultural centre, carrying on all the usual trades and industries of a town of that character. It was created a borough by Henry III and through the middle ages played its part in border and civil wars. In the days of the Commonwealth it had some strategic value in the possession of a bridge in the neighbourhood of Wilton Castle. The "Man of Ross," John Kyrle, a plain country gentleman but a great benefactor to the town, has acquired some fame above his very respectable merits and utility through Pope, who has immortalised him in verse. Ross has a fine high-perched church with a tall spire, and a beautiful seventeenth century town-hall. (pp. 16, 19, 21, 98, 110, 113, 118, 120, 121, 123, 126, 130.)

Stretton Sugwas (325), three miles north-west of Hereford, is the site of a former palace of the Bishops of Hereford. Its old Norman church has been recently rebuilt.

Titley (296), a large village on the Radnor border three miles north-east of Kington, where was formerly a priory. Wapley Hill, a well-known height and camp, is in the neighbourhood.

Walford (1025), on the Wye, two and a half miles south-south-west of Ross, was a noted ford into Wales (Walesford) and has a very interesting and ancient church and the celebrated black-and-white manor house of Wythall. The parliamentary headquarters were here during the siege of Goodrich Castle in the Civil War.

Welsh Bicknor (102) adjoins Symond's Yat on the southern border with its famous scenery, hotels, and lodging-houses, and is almost encircled by the Wye. It was formerly a detached portion of Monmouthshire. Henry V spent his childhood in the parish at Courtfield, the present seat of the Vaughan family.

Weobley (702), a large village, once a parliamentary borough, seven miles south-west of Leominster and the most visited by artists of all the Herefordshire half-timbered villages. There is an imposing fourteenth century church with a lofty spire, containing among other monuments one to Colonel Birch, a Cromwellian, who rose from an obscure position to fame in the Civil War, embellished the church, and acquired much property in the neighbourhood. There are traces of a castle here in Garnstone park. (pp. 108, 113, 121.)

Wigmore (387), seven miles south-west of Ludlow, has an Early English church but is chiefly noted for the ruins of the great and powerful castle of the Mortimers. (pp. 72, 73, 101, 116, 121.)

Woolhope (584), seven miles east-south-east of Hereford, has an interesting Early English church and a fine lych-gate, but is better known for the remarkable geological formation of the uplifted valley in which it lies. (pp. 30, 31.)

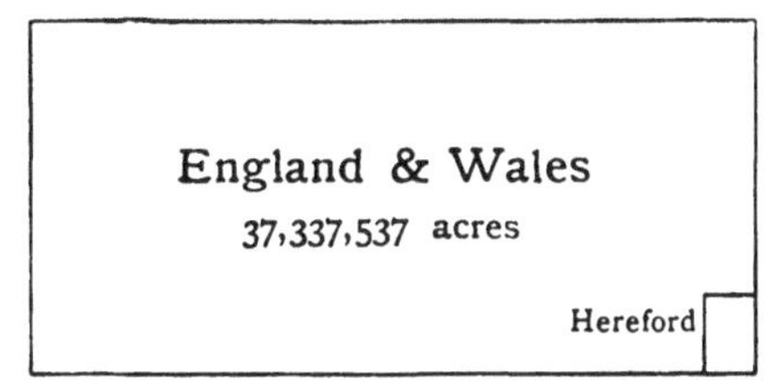

Fig. 1. The Area of Herefordshire (538,924 acres) compared with that of England and Wales

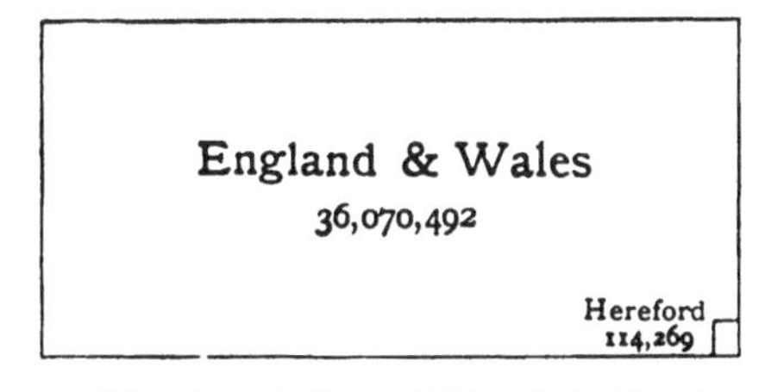

Fig. 2. The Population of Herefordshire (114,269) compared with that of England and Wales

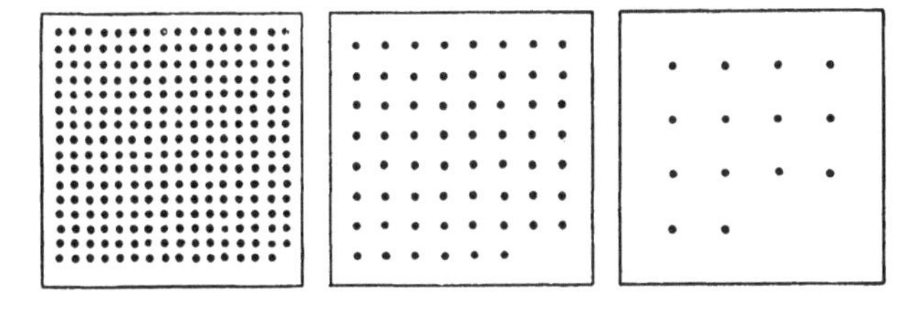

Lancashire 2554 England and Wales 618 Herefordshire 136

Fig. 3. Diagram showing comparative Density of Population to the Square Mile

(Each dot represents ten persons)

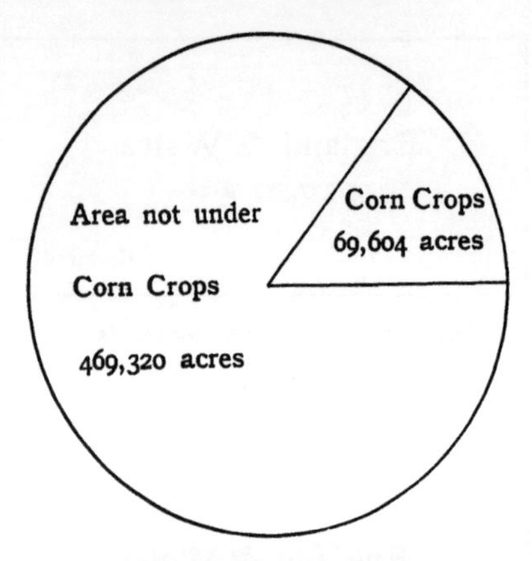

Fig. 4. Proportionate Area under Corn Crops in Herefordshire in 1911

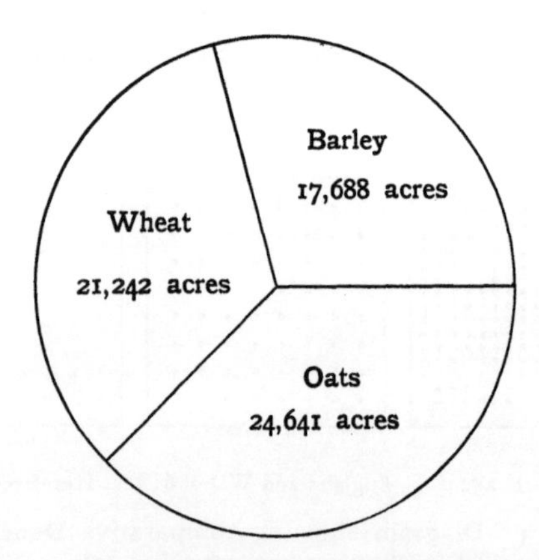

Fig. 5. Proportionate Area of chief Cereals in Herefordshire in 1911

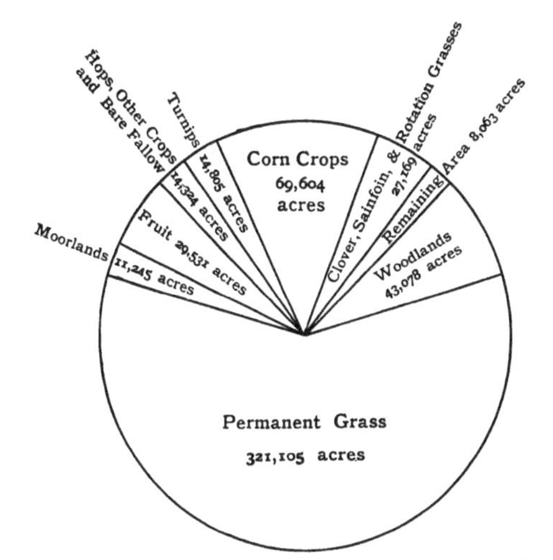

Fig. 6. Proportion of Permanent Grass to other Areas in Herefordshire in 1911

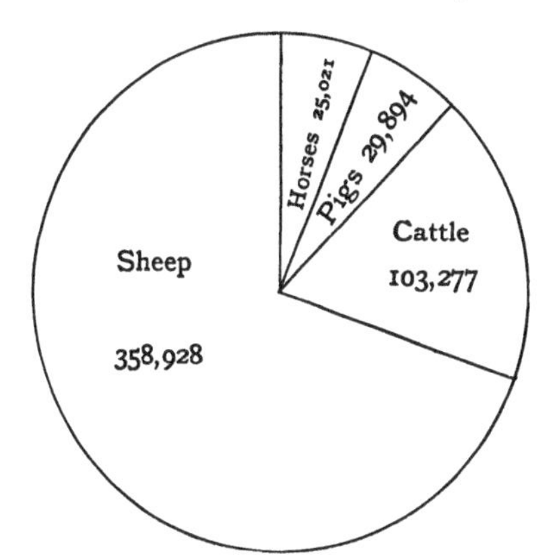

Fig. 7. Proportionate numbers of Live Stock in Herefordshire in 1911

www.ingramcontent.com/pod-product-compliance
Ingram Content Group UK Ltd.
Pitfield, Milton Keynes, MK11 3LW, UK
UKHW042145280225
455719UK00001B/112